PRAISE FOR *AGE OF THE CITY*

'A fresh, clear-eyed and timely analysis of the challenges and opportunities that comes from one of the most important themes of the 21st century – the rise of urbanisation and the fact that more people live in cities than at any time in human history.'

Professor Peter Frankopan, Oxford University and author of *The Earth Transformed* and *The Silk Roads*

'*Age of the City* takes us on an absorbing journey through the relationships connecting civilization, progress, and the city.'

Tim Marshall, author of *Prisoners of Geography*

'*Age of the City* is the book we need now. Ian Goldin and Tom Lee-Devlin take aim at those who believe the age of our great cities is over. They marshal powerful and much needed evidence to show that cities are becoming even more important to our economy and society. Their book illuminates the ongoing ability of cities to preserve and thrive in the face of all manner of adversity, as platforms to harness and unleash the human creativity which stands as the engine of human progress. **Their book is essential reading for political and business leaders and each and every one of us who cares about and wishes to help create a better collective future.'**

Professor Richard Florida, University of Toronto and author of *The Rise of the Creative Class*

'A sweeping survey of the history and modern challenges facing cities that will persuade you that they are the key to a happier and more sustainable future together.'

Baroness Minouche Shafik, President and Vice-Chancellor, London School of Economics and Political Science

'Ian Goldin and Tom Lee-Devlin have written a compelling volume explaining why cities will survive and thrive despite the twin threats of remote work and pandemic. This book vividly explains how cities are engines of cooperation, which fundamentally enable us to become more human. **Using a compelling combination of history and data, the authors remind us that life is better lived in urban streets and cafes than in Zoom waiting rooms. This is an important read** for anyone who cares about cities.'

Professor Edward Glaeser, Economics Department Chair, Harvard University and author of *Triumph of the City*

'**A compelling, holistic and well-balanced narrative** on the critical role of cities in an age of global warming – full of insights based on hard data. **From cover to cover, a great read.** Full of positive ideas for the future, and grounded in vital lessons from the past. The authors link together many disparate subjects into one integrated whole – bringing alive history, planning, infrastructure, pandemics, urbanism, deprivation, industrialisation, fertility, wars, governance and more – all in support of the city.'

Lord Norman Foster, architect and designer

'*Age of the City* provides a startlingly fresh and compellingly readable account of the forces that have defined our past and will shape our future. **An essential and enjoyable guide for all our lives.**'

Professor Saskia Sassen, Columbia University and author of *The Global City*

AGE OF THE CITY

Why Our Future Will Be Won Or Lost Together

IAN GOLDIN
TOM LEE-DEVLIN

BLOOMSBURY CONTINUUM
LONDON • OXFORD • NEW YORK • NEW DELHI • SYDNEY

BLOOMSBURY CONTINUUM
Bloomsbury Publishing Plc
50 Bedford Square, London, WC1B 3DP, UK
29 Earlsfort Terrace, Dublin 2, Ireland

BLOOMSBURY, BLOOMSBURY CONTINUUM and the Diana logo are trademarks of Bloomsbury Publishing Plc

First published in Great Britain 2023

A catalogue record for this book is available from the British Library

Library of Congress Cataloguing-in-Publication data has been applied for

ISBN: HB: 978-1-3994-0614-7; TPB: 978-1-3994-1242-1; eBook: 978-1-3994-0613-0; ePDF: 978-1-3994-0612-3

2 4 6 8 10 9 7 5 3 1

Typeset by Deanta Global Publishing Services, Chennai, India
Printed and bound in Great Britain by CPI Group (UK) Ltd, Croydon CR0 4YY

To Tess, for all your love and support
Ian Goldin

To Megan, for all the joy
Tom Lee-Devlin

ALSO BY IAN GOLDIN

Rescue: From Global Crisis to a Better World
Terra Incognita: 100 Maps to Survive the Next 100 Years
Age of Discovery: Navigating the Storms of Our Second Renaissance
Development: A Very Short Introduction
The Pursuit of Development: Economic Growth, Social Change and Ideas
Is the Planet Full?
The Butterfly Defect: How Globalization Creates Systemic Risks, and What to Do About It
Divided Nations: Why Global Governance is Failing, and What We Can Do About It
Exceptional People: How Migration Shaped our World and Will Define Our Future
Globalization for Development
The Case for Aid
The Economics of Sustainable Development
Economic Reform and Trade
Modelling Economy-wide Reforms
Trade Liberalization: Global Economic Implications
Open Economies
Trade: What's at Stake?
The Future of Agriculture
Lessons from Brazil
Making Race: The Politics and Economics of Racial Identity

CONTENTS

LIST OF FIGURES

PREFACE

We live in tumultuous times. In the space of just a few years, we have witnessed a surge in populist politics across the world, a global pandemic, a spike in environmental disasters and a fraying of geopolitical relations demonstrated by the tragic war in Ukraine and escalating tensions over Taiwan. That has all occurred against a backdrop of dramatic technological changes that are fundamentally altering the way we work and relate to one another.

So why do we need a book about cities? There are two reasons. First, cities are now home to over half of the global population, a share that will rise to two-thirds by 2050. That is something never before seen in human history, and means that the forces shaping life in cities now also shape our world as a whole. Second, cities throughout history have been the great incubators of human progress through their power to bring us closer together, something we need now more than ever.

Put another way, the battle for our future needs to be fought and won in cities. From inequality and seething social divisions to pandemics and climate change, this book will show how many of the answers to our greatest challenges are to be found in reforming our cities. It will also show that, if we fail to take action, cities will magnify the perils that lie ahead.

This is a book that has been many years in the making, with its origin in the great paradox of modern globalization: that declining friction in the movement of people, goods and information has made where you live more important than ever. Appreciation of the complexity of globalization has come a long way since the early 2000s, when Thomas Friedman's *The World is Flat* and Frances Cairncross's *The Death of Distance* captured the public's imagination. With the power of hindsight, we now know that place is of utmost importance in a globalized world.

That world is being shaken by a populist politics, often built on anger against cosmopolitan urban elites in major global cities. This has been given expression through Brexit in Britain, and in support for anti-establishment politicians in the US, France, Italy, Sweden and other countries. A common thread of all these populist movements is the notion that mainstream politicians, business leaders and media figures cocooned in big cities have let the rest of their countries down and lost interest in 'left behind' places and people. These populist revolts against dynamic cities are rooted in real grievances based on stagnating wages and soaring inequality. A transformational effort to even out economic opportunity is long overdue. But undermining dynamic cities is not the way to do that. Cities like London, New York or Paris – and in the developing world Mumbai, Cairo or Lagos – are engines of economic growth and job creation without which their respective national economies would be crippled. Moreover, they continue to harbour profound inequalities of their own, which need urgent redress. That is why we advocate in this book for a holistic approach to economic revival that harnesses the power of cities, rather than trying to resist it.

The impact of the recent surge in remote working on the geography of our economy also demands answers, which this book seeks to provide. Without a doubt, the collapse of commitment to offices and commuting is proving to be highly disruptive for cities, particularly in the US and Europe where rates of remote work remain high. Commercial real estate is suffering, municipal taxes are declining and the viability of businesses that depend on intense footfall – from barbers to buskers – is being challenged. So too are public transport systems, many of which are haemorrhaging cash.

All of that is reason to rethink cities, not abandon them. Creativity still thrives on physical interactions and serendipitous encounters. Most jobs are still apprenticeships, meaning workers, especially early on in their careers, benefit from observing and informally engaging with their more experienced colleagues. And workplaces are pillars of the community, bringing together people from many different walks of life and helping to combat isolation and loneliness. A society without dynamic cities would be less productive, less cohesive and less fulfilled. Our argument in this book is that, with the right initiative, the transition to hybrid work offers a window of opportunity to transform cities for the better.

While much of this book focuses on the cities of the developed world, we also offer readers a global perspective. The growth in the share of the world's population living in cities in recent decades has been driven almost entirely by developing countries, which now account for most of the world's urban-dwelling population. In some of those countries, such as China, rapid urbanization has been the result of a process of economic modernization that has lifted large swathes of the population out of poverty. In others, such as the Democratic Republic of the Congo, urbanization and economic development have been disconnected, with rural deprivation and the flight from danger playing a greater role in the migration to cities than urban opportunity. Either way, cities are now where the world's poor are choosing to live. And many of their cities are giant and overcrowded, with residents too often living in appalling conditions.

Appreciating what is happening in the cities of the developing world is essential if poverty is to be overcome. It also is vital if we are to understand why contagious diseases are making a comeback. Modern pandemics, from HIV to Covid-19, have their origins in these cities. Crowded conditions are coinciding with a number of other trends in poor countries, including rapid deforestation, intensive livestock farming and the consumption of bushmeat, to increase the risk of diseases transferring from animals to humans and gaining a foothold in the population. From there, connectivity between the world's cities, particularly via airports, makes them a catalyst for the global dissemination of deadly diseases. That means that dreadful living conditions in many developing world cities are not only a humanitarian and development issue, but also a matter of global public health.

Tremendous progress has been made in the past two centuries in combating infectious diseases, but the tide is turning against us. Cities will be the principal battleground for the fight ahead.

No book on the challenges currently confronting the world would be complete without consideration of climate change. This poses an existential threat for many of the world's cities. Ocean rise, depletion of vital water resources, and urban heatwaves risk making many cities uninhabitable. Coastal cities are particularly vulnerable, yet nearly all global urban growth is in coastal cities. While rich cities such as Miami, Dubai and Amsterdam are threatened, those in poorer countries are even more vulnerable as the monumental investments required to build sea walls and drainage systems are simply unaffordable. Cities nevertheless hold many of the answers to mitigating climate change, and we show how they can establish more sustainable foundations.

In this book, we bring together insights from a wide range of disciplines in order to inform our understanding of the challenges facing cities and their potential. Historians, economists, sociologists, urban planners and other experts all look at cities through different lenses. Each are valuable, but problems do not emerge in disciplinary silos, nor do solutions.

We are far from the first to recognize the fundamental importance of cities to the modern world. Ed Glaeser's *Triumph of the City*, Richard Florida's *The Rise of the Creative Class*, Enrico Moretti's *The New Geography of Jobs* and many other excellent books over recent years have laid a trail before us, as have canonical works such as Lewis Mumford's *The City in History*, Peter Hall's *Cities in Civilization*, Jane Jacobs' *The Death and Life of Great American Cities* and Paul Bairoch's *Cities and Economic Development*. We have also been inspired by recent works that explore the importance of place more broadly, such as Tim Marshall's *Prisoners of Geography*, and provide a fresh historical perspective on why our world is the way it is today, including Peter Frankopan's *The Silk Roads*. We hope this book can further contribute to advancing our collective knowledge on these subjects.

This book argues that we must take action now to shape our urban destiny, and shows how the challenges we face should be addressed. By sharing our understanding of the present and vision for the future, our aim is to equip our readers to play a part in creating a better life for all.

Ian Goldin, Oxford
Tom Lee-Devlin, London

1

Introduction

With the birth of the city roughly 5,000 years ago, our ancestors began a journey that would forever change our relationship with the world around us and with each other. Cities, by bringing people together, unleashed the potential for humankind to be more than the sum of its parts.

For most of the past, very few humans lived in cities. The Roman Empire, which achieved a historically unprecedented level of urbanization, never saw more than around 15 per cent of its population living in cities. Most other civilizations fell well short of that. Access to surplus food was the limiting factor. People who live in cities have always been fed by people who live in the countryside, and for most of the period since the dawn of agriculture rural populations have had little food to spare. Whenever food production increased, mortality rates fell, increasing the number of mouths to feed, eventually eliminating the surplus. The result was a ceiling on the share of people who could do anything other than work the land.

Yet profound structural shifts were also underway, as cities nurtured the potential for individuals to cooperate, specialize and invent at scale. Without those powers, *Homo sapiens* never would have been able to escape from its perilous state of subsistence.

In recent centuries, the previously slow and erratic evolution of human progress has accelerated. In an extraordinarily short period, relative to our roughly 200,000 years of existence, humanity has become many times more prosperous. Cities, as the catalysts for both industrialization and globalization, have been central to this progress. And the process has been self-sustaining. Innovations in food production, storage, processing and transport have meant that the global agricultural system has been able to sustain rapid urban growth. Advances in public health, sanitation and infrastructure have allowed far greater numbers of people to concentrate together geographically than was ever previously possible.

In the early years of the twenty-first century, cities for the first time became home to most of humanity. At the beginning of the eighteenth century, just 5 per cent of the world's population lived in cities.[1] Today, that share is 55 per cent.[2] By 2050 it is predicted that over two-thirds of the world's population will be urban. Figure 1 illustrates the accelerating speed of this transition. As recently as 1960, only 40 countries had a majority urban population; by 2020, more than 100 countries had passed that threshold.[3] Between today and 2050 the global urban population will increase by at least 2.5 billion people, with most of that growth occurring in the developing world as continued migration from the countryside to cities combines with rapid population growth.[4] In the next ten years, more than two dozen new cities will cross the threshold of five million inhabitants.[5] *Homo sapiens* may have evolved on the savannah, but we are now an urban species.

As we become increasingly urban, we face a multitude of threats that are taking on an existential nature. These include widening inequality, fraying social bonds and trust, and rising systemic risks, particularly from pandemics and climate change. These must be resolved if we are to avoid a dystopian future. Only by building fairer, more cohesive and more sustainable cities will we be able to address these challenges.

FIGURE 1: More than half the world's population now lives in cities

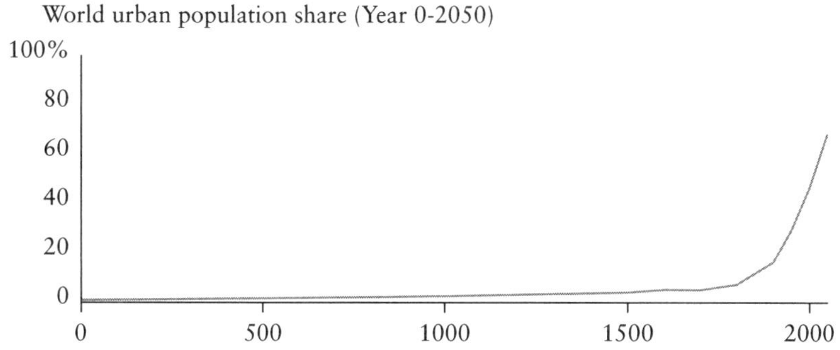

Source: H. Ritchie and M. Roser, *Urbanization (Our World in Data)*, 2019.

THE GREAT EXPANSION

From the Griffith Observatory one can start to get a perspective of the scale of modern-day Los Angeles, stretching from the San Gabriel Mountains to the Pacific Ocean. After dark, the lights of 13 million inhabitants shine brighter than the night sky.

The modern city is unlike anything that has come before it. Historically, a city with over one million inhabitants was confined to rare examples like Ancient Rome in the second century CE or Chang'an (modern-day Xi'an) in the eighth century CE. Renaissance Florence, despite its wealth and intellectual fervour, had a population of about 100,000. Cities through most of history did not expand beyond easy walking distance.[6] London did not grow beyond the original 'square mile' until the seventeenth century. And while the physical extent of the city was curtailed by a lack of modern transportation, its population density was curtailed by the elevated risk of infectious disease that came with crowding.

The result was that new urban areas were formed once existing ones had outgrown their manageable scale. Early in the development of Ancient Greece, for example, the Delphic oracle would regularly order the dispersal of cities that grew to an unwieldy scale, giving shape to the patchwork of city-states that defined that civilization.[7] The early settlers of New England followed a similar model.[8]

Throughout history, many argued that cities were an unnatural state for humanity, a culmination of our exile from the Garden of Eden. The influential theologian St Augustine presented the human story as a conflict between the iniquitous 'City of Man' and the heavenly 'City of God', observing that it was Cain, murderer of his brother Abel, who according to the biblical narrative founded the first human city.[9] Later, Jean-Jacques Rousseau would declare cities to be 'the abyss of the human species',[10] while Thomas Jefferson would dismiss them as 'pestilential to the morals, the health and the liberties of man'.[11] The essayist William Hazlitt feared that the city engendered 'a puny, sickly, unwholesome, and degenerate race of beings'.[12]

Nevertheless, cities began to grow rapidly in size following the onset of the industrial revolution in Europe in the eighteenth century. A succession of advances in agriculture, from crop rotation to the mechanical reaper, created a much larger agricultural surplus in the countries where they were adopted. In Britain, the enclosure movement, under which lands that were previously communal and accessible to all were appropriated for private landowners, displaced a sizeable share of the rural population. At the same time, the adoption of new manufacturing technologies, including the spinning-jenny and steam engine, created enormous numbers of jobs in booming urban industries such as textiles and metalworks.

As populations flooded into early industrial cities such as Manchester, they quickly became crowded and unsanitary. With the invention of the railway, it became possible to move many more goods and heavier loads over longer distances far more quickly. That in turn allowed industrial activity to take root in a greater number of cities, but the dispersal was not quick enough to outpace the continued mass migration into urban areas. As a result, cities continued to grow ever larger, with lower mortality rates thanks to the advancing state of medicine and public health adding to this growth.

Ebenezer Howard was one of the many who had grown appalled by the living conditions of the poor in industrial cities. Working as a parliamentary reporter at the end of the nineteenth century, he developed a passion for urban planning, and drafted a proposal for what he called the 'Garden City'. His vision was to disperse the populations of overcrowded cities into a constellation of connected but self-sufficient towns containing a mixture of spacious housing, public amenities, some factories and an agricultural belt around the periphery.[13] Other than a few isolated cases, such as Letchworth and Welwyn on the outskirts of London, Howard's model was never adopted in its entirety, but his ideas profoundly influenced urban planning over the course of the twentieth century. The new principle of urban design was to offer people the closest thing possible to the spacious lifestyle of the countryside while maintaining access to the economic opportunities afforded by a city. And so was born the idea of suburbia.

While the streetcar and intra-city rail made some degree of dispersion possible in the early decades of the twentieth century, suburbanization didn't really begin in earnest until the middle of the century when cars became widely accessible for the growing middle class. Once they did, the change was rapid. Consider the case of New York. Since 1950, the population of the wider metropolitan area has risen from 13 million to 20 million, but the population of the five central boroughs is still roughly what it was 70 years ago, around 8 million.[14] Much has been made of the revival of New York's inner city neighbourhoods in recent decades from their nadir in the 1980s, but that has not stemmed the growth of its suburbs.

The city, in other words, has burst its seams. Today, there are more than 500 urban areas with over 1 million residents and 40 with over 10 million.[15] In extreme cases, the process of sprawl has led to adjacent cities merging into single urban agglomerations. One such example is the melding together of Tokyo, Yokohama, Chiba and a number of other nearby cities into Greater Tokyo, the largest city in the world with a population of 37 million. Back in 2008, the urbanist Richard Florida and colleagues identified 40 such 'mega regions' around the world that together accounted for two-thirds of global economic output.[16] The trend has further accelerated since then. The rapidly growing Pearl River Delta area in China, for example, encompasses nine increasingly interconnected cities with a total population of 65 million, equivalent to that of the UK or France, and a total output of $1.2 trillion, nearly equivalent to that of Spain.[17]

In his 1961 magnum opus *The City in History*, renowned historian Lewis Mumford described the city as both a container and a magnet.[18] As cities have continued to bleed outwards, it is the magnet that has become the most apt metaphor for understanding our urban world. It may be difficult to draw a neat boundary around the city today, but its gravitational pull is more powerful than ever.

The recent surge in remote working has led some to imagine that the city may soon become a relic of a past era in which humanity was held hostage by the need for physical proximity. Untethered by this constraint, we will be able to tidily disperse ourselves across the planet like gas molecules. Despite their outward expansion, cities still represent just 1 per cent of earth's habitable landmass.[19,20] If we can work from anywhere, why should we continue to make such little use of the bountiful space available to us?

Cities are, however, more than just office parks, and provide their inhabitants with a wide range of services that can only be delivered in person. There is no virtual solution for a sports massage or childcare, and anyone who experienced a virtual cocktail party during the pandemic will share our scepticism about such alternatives to real-world hospitality. Most people cannot do their jobs remotely. It is an option for only about one in three workers in rich countries, and even fewer in poor countries.[21]

Moreover, cities have a crucial role to play in building cohesion in a world in which our social interactions are dominated by the internet, funnelling us into echo chambers full of people who think like us and undermining our ability to find common ground with those who think otherwise. The cities of our ancestors played a vital role in bringing together people from different walks of life and binding them into a cohesive whole. In our fractured world, cities must rediscover this potential. That will require them to evolve. Low-density sprawl, which trades connection for privacy, is one-half of that challenge. The other is the rising inequality – both within and between cities – that has led the rich and the poor to inhabit largely separate worlds.

THE WORLD IS NOT FLAT

The years between the fall of the Berlin Wall and the global financial crisis were days of heady excitement over globalization. A wave of trade deals, capped off in 2001 by China's entry into the World Trade Organization, integrated the world's major economies. Many poor countries appeared to finally be catching up economically on the back of increased trade and soaring demand for their commodities. And the internet had arrived.

The development of the World Wide Web in 1989 was followed by surging investment in communications infrastructure. By 2005 more than 1 billion people were connected over the internet.[22] That year, the award-winning *New York Times* journalist Thomas Friedman published a daringly titled book, *The World is Flat*, in which he made the case that the global economic playing field was rapidly levelling. An instant bestseller, it captured the optimism of the era.

But while the world was becoming increasingly frictionless, thanks to the liberalization of trade and the wonders of the internet, another profound change was also at work, the result of which has been an uneven world, more spiky than flat, in which place is more important than ever. The combination of globalization and rapid technological progress has led to the disappearance of many manufacturing and clerical jobs – long anchors of the middle class – as work was either eliminated entirely or sent overseas to be performed more cheaply. As these jobs have disappeared, a barbell-shaped economy has emerged, characterized by high-paid knowledge jobs, such as management consultants and data scientists, on the one hand and low-paid service jobs, like baristas and warehouse workers, on the other. In the words of economists Maarten Goos and Alan Manning, we have seen the workforce split into 'lousy' and 'lovely' jobs.[23]

Manufacturing jobs in particular have long been the life force behind smaller cities and also many rural towns. Between the early nineteenth and late twentieth centuries, canals, then railways, then highways led to a dispersion of manufacturing across the United States, Britain and other countries. When these jobs disappeared, many of the places that relied on them struggled to adapt. Public finances became stretched, infrastructure decayed, crime rose and health outcomes deteriorated. In recent years, these left-behind places have increasingly become poverty traps for their residents.

Meanwhile, a small number of leading cities have never had it so good. Cities such as New York, London, Paris, Shanghai and Tokyo are the gateways of the global economy. They not only serve as physical conduits to trade via seaports and airports, but are also host to the multinational offices, financial institutions and professional services firms that make international flows possible. The size of these cities, combined with the frequent presence of top universities, makes them attractive to workers in fields such as computer science or biomedical research where the cross-pollination and dissemination of ideas underpin performance. At the other end of the job scale, those in lower-paid service occupations such as hospitality, childcare, fitness and retail are also flocking to these cities as the presence of high-skill workers with disposable incomes drives demand for their services.

A century ago, German historian Oswald Spengler published his magisterial work *The Decline of the West*. In it he predicted that the division between 'cosmopolitans' and 'provincials' would eventually become the defining contrast in modern society, dominating 'all events, all habits of life, all views of the world'.[24] Recent history has proved that to be an eerily prescient forecast. Today's global cities are increasingly connected with one another and increasingly disconnected with struggling places in their own countries. A banker in New York might travel to London multiple times a year – at least before the pandemic – without ever having set foot in a city like Baltimore or spent more than a fleeting moment in a country town. The picture of a prosperous cosmopolitan elite with little interest in the plight of their fellow citizens has been a rallying motif for an anti-globalist populism across the United States, Britain, France and many other rich countries, fuelled by deep resentment. The fact that many educated elites in London knew no one who had voted for Brexit, or in New York who voted for Trump, was the ultimate testament to how disconnected they had become from their wider societies.

While the gulf *between* thriving cities and everywhere else has widened, so too has the disparity *within* these cities. Inequality has risen in most metropolitan areas in the United States since 1980, but it has risen fastest in large, thriving cities like New York, San Francisco and Chicago, where it is now much higher than the national average.[25] Wages for high-skilled knowledge workers in these places have soared, but pay for the low-skilled service workers supporting them has languished, a gulf that is compounded by the rapidly rising cost of living in these cities. As knowledge workers have flocked to gentrifying inner cities in recent decades, these global metropolises have increasingly come to resemble ivory towers, with a highly concentrated core of prosperity served by a sprawling periphery of disadvantage.

The rising concentration of wealth and opportunity in a small number of cities and a small number of neighbourhoods within them is not sustainable. It is not just deeply unfair, but also a massive economic failure, as large swathes of the population in rich countries are prevented from reaching their full productive potential by the lottery of where they are born. This challenge is not insurmountable. In this book, we will explore how cities around the world – from Seattle to Leipzig to Fukuoka – have managed to catch up with thriving metropolises, and the lessons they hold for cities that are struggling today. And we will show how, with the right reforms to education, housing and public transport, cities can create economic opportunity for all their residents, not just a lucky few.

It is not only rich countries that are being affected by technological change. In the post-war decades, economies like Japan and South Korea were able to follow a path to prosperity that was built on their ability to leverage their lower labour costs to export cheap manufactures to rich countries. A similar model was followed by China after it began liberalizing its economy. For these countries, globalization has been their ticket out of poverty, transforming them from rural agricultural economies to urban manufacturing centres in a matter of decades. However, as automation undermines the economic advantage provided by cheap labour, it looks increasingly unlikely that economies that are still early along the development path will be able to follow past labour-intensive models of growth. We will explore what the future of economic development might look like for these countries, as well as the consequences if they fail to create a sufficient number of good jobs for their growing populations.

THE PERILS OF THE ANTHROPOCENE

Cities – by unleashing humankind's potential for cooperation, specialization and invention – have made us masters of the earth. Yet our fate is still fundamentally tethered to the natural world, and our actions are putting that relationship in danger.

The first cities did not emerge until a period of relative environmental stability for our planet known as the Holocene. A mild and stable environment created the necessary conditions for *Homo sapiens* to transition from hunting and gathering to agriculture. With time, we became more productive farmers, which eventually made it possible for the first urban dwellers to untether themselves from toiling the land and gather together in cities. Now, the Holocene is being replaced by the Anthropocene, as a growing population and rising living standards are together straining the limits of our planet and undermining the foundation upon which the success of our species has thus far depended.

It is beyond dispute that human-induced climate change is accelerating. In the last 50 years alone the global average surface temperature has risen by about 1 °C,[26] the Arctic sea ice minimum has shrunk by 35 to 40 per cent,[27] and the frequency of natural disasters has increased five-fold.[28] We are simply doing too little too late to reduce greenhouse gas emissions.

Cities will be hit hard by climate change in the decades ahead. Many major cities are situated next to oceans or rivers or both. Major coastal cities such as Miami, Los Angeles, Jakarta and Shanghai risk being inundated by rising water levels. Others face droughts, water shortages and intolerable extremes of heat. In the developing world, falling agricultural yields will further accelerate rural–urban migration, pushing the already-strained infrastructure of cities like Lagos and Dhaka beyond breaking point. A humanitarian crisis of immense proportions is on the horizon unless cities ready themselves for the challenges ahead. The effects of failing to act will be felt around the world. Accepting one million Syrian refugees destabilized European politics in a way few expected. Migration as a result of climate crises will be orders of magnitude greater.

Cities are also essential to halting climate change. This is partly because they generate 70 per cent of the world's carbon emissions, but also because dense urban living is far less emissions-intensive than life in the countryside. Yet urban emissions remain far too high to prevent runaway climate change.

The good news is that many cities are already taking action to reduce emissions further, by making public transport more accessible, encouraging the use of electric vehicles, improving the energy efficiency of buildings and adding green spaces that not only lower urban temperatures but also act as carbon sinks. Still, there is much more that needs to be done, including discouraging car-based suburban sprawl and tackling emissions generated beyond the city by reducing waste and improving recycling rates.

It is also vital to honour the 2015 commitment made to developing countries to provide more than $100 billion a year to help finance a less emissions-intensive path to economic development and build more sustainable cities to house their rapidly growing urban populations. If the fast-growing cities of the developing world lock in the same types of energy-consumption patterns as cities like Houston, there will be no hope of stopping runaway climate change.

The experience of the Covid-19 pandemic has also vividly demonstrated that climate change is not the only existential threat we face in the decades ahead. In the past 200 years advances in medicine and public health have allowed humanity to gain the upper hand over infectious diseases, helping to dramatically improve mortality rates, especially in dense cities, and fuelling a boom in the global population. Now, our increasing density and interconnectedness is colliding with a host of other factors – including the intensification of livestock farming, the disruption of wild habitats through deforestation, and the rising risk of antimicrobial resistance – to turn the tide back in favour of infectious diseases.

The Covid-19 pandemic has led to the tragic loss of over 20 million lives around the world,[29] yet history shows that far worse is possible. It would be profoundly naïve to assume that science will be able to indefinitely forestall the emergence of even more lethal pandemics. It is imperative that we learn from

Covid-19 how to better handle future pandemics when they inevitably arise. The density of urban living makes cities natural amplifiers of outbreaks, while their role as the connecting nodes across borders also makes them an essential first line of defence in containing the global spread of diseases. Cities therefore must form a central pillar of any strategy to better protect us from pandemics going forward.

A GUIDE TO THIS BOOK

Following this introduction, in Chapter 2 we show how cities have played a pivotal role in spurring human progress from Ancient Mesopotamia to today. In Chapter 3, we unpack how the recent transition to a globalized knowledge economy has consolidated wealth into a small number of leading major cities, and consider the actions needed to 'level up' places that are falling behind. In Chapter 4, we turn our attention to why poverty and riches coexist within cities, and set out what needs to be done to ensure cities offer the opportunity of a better life for all their citizens.

We then turn our attention to a deeper exploration of the role of cities as centres of work and community, with a particular focus on the impact of the internet. In Chapter 5, we examine the opportunities and limitations of remote working, and its implications for cities. In Chapter 6, we explore how and why the internet is accelerating social fragmentation, and what cities can do to help reverse this trend.

We next widen our aperture, in Chapter 7, to consider cities in poorer countries, including the unique challenges that arise from the speed with which they are growing, and the risk of urbanization without prosperity as the traditional manufacturing-led development path becomes increasingly difficult to follow in light of the accelerating pace of automation.

Finally, we consider two existential risks confronting humanity – infectious diseases and climate change – and what they mean for cities. In Chapter 8, we look back at the relationship between infectious diseases and cities through history, reflect on the learnings from the coronavirus pandemic and define how cities can help make us more resilient against future pandemics. In Chapter 9, we explore the potential disruption that cities will face from climate change in the decades ahead, the actions that need to be taken to protect populations from the rising impact and frequency of flood, drought and heat extremes, and the role that cities must play in helping to accelerate decarbonization and slow climate change.

We conclude the book, in Chapter 10, with a summary of the agenda required to ensure cities rise to their potential and contribute to a fairer, more cohesive and more sustainable world. If humanity is to ensure that the coming age is our best yet, we must overhaul urban design, rebuild for the knowledge economy and accelerate sustainable development. That is what we owe future generations.

AN ASIDE ON DEFINING CITIES[30]

There is no universal agreement on what is meant by a city. For some it is defined by population size or density. For others it is an administrative designation. In England, having a cathedral helps places to qualify, so Ian's home town, Oxford, with a population of around 160,000 is defined as a city, whereas many towns that are larger are not. In the absence of any internationally agreed definition, there could be anywhere between 10,000 and a million cities in the world. One challenge is delineating boundaries for sprawling areas, with towns and cities combining into massive metropolitan regions. Distinguishing between urban and rural areas is similarly far from trivial. In Iceland, any community of over 200 people is defined as urban, in Canada or Australia the threshold is 1,000, in Portugal 10,000 and in Japan 30,000. In China, places with at least 1,500 people per square kilometre are designated urban. Atlanta, with only 600 people per square kilometre, would not qualify. Las Vegas, with 1,975 residents per square kilometre, would just scrape through.

Rather than get caught up in a fruitless attempt to reconcile these definitions, our approach is to avoid the semantic disagreements and to focus our attention on the phenomenon of rapid urban growth. The trend towards ever-bigger clusters of adjacent people and buildings is indisputable. For us, cities are best thought of as physical, social, economic and political communities where individuals live, work and connect with one another. We trust our readers know a city when they see it, and do not need us to labour the definition.

2

Engines of Progress

Cities have always been the engines of human progress. From the birth of civilization in the cities of Mesopotamia to the first flowerings of democracy in Ancient Athens, from the intellectual flourishing of Renaissance Florence to the first spark of the industrial revolution in nineteenth-century Manchester, cities have been the epicentres of humanity's greatest achievements.

If the entire story of *Homo sapiens* was compressed into one day, the first city would not have emerged until after 11 p.m. For most of our history we were principally a nomadic species. With the rise of agriculture, the first small human settlements began to emerge between 10,000 and 12,000 years ago.[1] As humans learnt to store food and domesticate animals, they began to settle down in small communities. With time, those settlements grew and became more complex. The fertile floodplains of southern Mesopotamia, in modern-day Iraq, were home to a number of these growing settlements. Around 3500 BCE, the most famous of these – Uruk – passed a threshold of scale and sophistication for what one might consider a city.[2] The archaeological record of Uruk points to a human society with written language, organized religion, skilled feats of engineering and a clear sense of itself as a united community.[3] With the first cities came the birth of civilization, a fact reflected in the common Latin root that underpins these two words.[4]

Cities – and civilizations – emerged independently of one another in different regions of the world within a relatively short period of time, at least compared to the very slow evolution that preceded it. By 3100 BCE, a distinct civilization had emerged in Egypt, centred on the city of Memphis.[5] The Indus Valley civilization, centred on the ancient cities of Harappa and Mohenjo-Daro on the Indus River in modern-day Pakistan, emerged around 2600 BCE.[6] By around 1900 BCE, Chinese civilization had emerged in the Yellow River region, home to a number of ancient cities including Luoyang, Anyang and Chang'an.[7] And by around 400 BCE, Mesoamerican civilization had taken shape around the city-states of the Maya.[8]

To understand why cities emerged, it is helpful to look back at the earlier transition of *Homo sapiens* from hunting and gathering to agriculture. While global population growth in the era before agriculture was a fledgling 0.01 to 0.03 per cent per year, when compounded over millennia the result was a total human population approaching 10 million at the cusp of the agricultural revolution, pushing against the upper boundaries of what could be supported by hunting and gathering.[9] Support for the role of population growth in the transition to agriculture can be found as far back as the ancient Chinese legend of the 'divine husbandman' Shen Nung: 'The people of old ate the meat of animals and birds. But in the time of Shen Nung, there were so many people that there were no longer enough animals and birds to supply their needs. So it was that Shen Nung taught the people how to cultivate the earth.'[10]

This process was helped by the ending around 12,000 years ago of the last great ice age and the transition to the period referred to by geologists as the Holocene. Warmer temperatures and an increase in air moisture provided ideal conditions for cultivated plants like wheat and rice to flourish.[11] Rising water levels from glacial melting also created a number of fertile floodplains that would come to play a foundational role in the birth of cities.

Environmental conditions continued to improve over the coming millennia, and by 5000 BCE flooding and sea levels had largely stabilized. The result was that river areas like those in Mesopotamia, the Nile delta, the Indus Valley and the Yellow

River region were able to support a far greater population density than previously was possible.[12] The unusually high fertility of the alluvial soil also facilitated agricultural surpluses, which in turn permitted a meaningful number of inhabitants to undertake activities other than the cultivation of land. Over time, the improved availability of food increased rates of reproduction and survival, and more individuals had the option of moving to urban areas where they could live less arduous lives than in the countryside.

This new potential to come together in large numbers set in motion a self-reinforcing process of growth that would forever transform the way humanity lives. While much of human progress – whether measured by population growth, material production or life expectancy – has taken place in the short period of time since the beginning of the industrial revolution, cities throughout recorded history have been laying the necessary foundations for that historic leap forward.

Behind this process lie three interconnected forces, each unleashed by cities. These are cooperation, specialization and invention, and together they have shaped the human story.

COOPERATION

What makes humans so successful as a species is our ability to operate together in large numbers in pursuit of common objectives. The late evolutionary biologist Edward Osborne Wilson argued forcefully against the belief that seemingly altruistic human acts merely reflect our 'selfish genes' protecting those who share our genetic material.[13] Instead, he saw *Homo sapiens* as a 'eusocial' species that acts in complex harmony in the interests of the group, similar to ants or bees.

However, it was not until the emergence of the city that *Homo sapiens* developed the mechanisms needed to facilitate cooperation beyond the scale of a small tribe constantly on the move. For its first two hundred millennia, humankind lacked a hive, and was able to coordinate itself in bands of probably no more than 150 fellow humans.[14] It is in cities that the art of human cooperation at scale was born.

As population densities continued to rise in the fertile regions in which the first cities emerged, new approaches to agriculture were required to feed the growing numbers of inhabitants. As nutrition improved, so did life expectancy. Population growth made it imperative to find ways to make the land more productive. Among the most important was the adoption of irrigation systems that drew water from rivers, which reduced reliance on unpredictable rainfall and greatly increased the carrying capacity of riverine cities. Building irrigation and other infrastructure systems was not something a single family could achieve by itself. It required working in cooperation at a scale that had never been demanded of previous generations.

In addition to irrigation, the populations of these fertile areas had to come together to protect themselves from the periodic flooding that threatens many river basins. The fact that cataclysmic floods feature in so many ancient mythologies – from Mesopotamia to Mesoamerica – is no accident. Such events were no doubt a constant fear for the inhabitants of early civilizations. In China, one can still find temples dedicated to Yu the Great, a hero of prehistory who introduced flood control to the Yellow River. Legend has it that Yu had two helpers in this endeavour: the Yellow Dragon, who used his long tail to carve out channels for flood water, and the Black Turtle, who used his flippers to build great levees out of mud.[15]

Engineering feats such as these were a major undertaking for ancient civilizations, requiring a joint effort across the community. In Uruk, and in many other ancient cities, it was religion that first performed this organizing role, helping to forge a sense of shared identity and mutual obligation.[16] Long before there were palaces and kings, there were temples and priests. In Uruk's case, the priestly caste eventually began to play a broader role in coordinating the affairs of the inhabitants of the region.[17] Indeed, it is in the temples of Uruk where writing is believed to have first appeared, as a method of bookkeeping. The Kushim Tablet is the oldest known record of written language, recording a transaction of barley that took place in ancient Uruk.[18] From these humble beginnings came one of the most important foundations of civilization, and the start of recorded history.

It was not only temples that brought solidarity to the inhabitants of ancient cities. Those who lived in Uruk and the other early cities would regularly encounter the same people in the streets or at the temple or marketplace. They would share stories, exchange information and probably gossip. Occasionally they would fight, and be forced to reconcile, or at the very least to tolerate each other. In prehistory, it was the ties of family – by blood or by marriage – that served as the glue that maintained mutual commitment. By bringing people together into close proximity, cities made it possible to begin thinking not just of the needs of one's family, but also of a much larger community.

While temple priests acted as the original authorities in early cities like Uruk, power eventually passed from priesthood to king.[19] Uruk may have been the first city to develop in southern Mesopotamia, but a number of others, such as Ur, Kish and Lagash, soon emerged in the surrounding area. With agricultural surplus came the temptation of conquest and the imperative of defence. Security became an essential foundation for survival, with a more centralized model of power needed to mobilize citizens for defence and armed combat.

The combination of an expansion in the agricultural surplus and a centralization of power over that surplus brought an end to the relative egalitarianism that had prevailed since the first days of hunting and gathering.[20] A chieftain may have been able to command their tribe, but with very little food to go around there was limited scope for them to wield that power to achieve a vastly superior standard of living for themselves. But kings, through a combination of charisma, violence and a network of patronage, could sustain a very different life from their subjects.

The *Epic of Gilgamesh*, the first known piece of literature, tells a mythologized tale of the adventures of King Gilgamesh, who reigned over Uruk during the third millennium BCE. The narrative follows a sequence of trials and tribulations in which Gilgamesh battles with gods and nature and seeks out the source of eternal life. Along the way, the tale provides a glimpse into the ambivalent feelings of the people of Uruk towards their ruler, who is portrayed as both a powerful protector and a greedy tyrant who 'lets no girl go free to her bridegroom'.[21]

Nevertheless, with a few exceptions, kingship emerged in some form or another in every ancient civilization, demonstrating the importance of both consent and coercion in sustaining cooperation in early societies. Not until the fifth century BCE, during the golden age of Athens, do we see the first glimmers of a viable alternative model. In the great experiment of Athenian democracy, every (male) citizen was made equal before the law and empowered to participate in collective debate over the governance of their society. For all its brilliance, the Athenian system was also deeply flawed. Citizens represented fewer than 20 per cent of the population, and were supported by a much larger population of resident foreigners and slaves, who, like women, lacked political representation.[22,23] Athens' treatment of foreigners reflects an all-too-common human tendency to draw boundaries and treat those who fall outside them by a different standard. Socrates may have declared himself a 'citizen of the world', but Athens as a whole was not above invading its rivals and enslaving their people.

In the ancient world, the two main ways in which a community could increase its surplus were to increase the amount of food produced per hectare of land or increase the number of hectares under control. The drive for prosperity led city-states like those of Mesopotamia to wage war in a bid to increase their land and subjugate their rivals.

Eventually, this process began to turn city-states into empires, with a dominant capital ruling over a network of tributary cities. The Old Kingdom of Egypt, established around 2700 BCE, was probably the first integrated society comprised of multiple urban centres, with a single pharaoh in Memphis ruling over much of the fertile land around the River Nile.[24] It was not, however, until later that the scale of these societies grew into what we would typically think of as an empire. By 500 BCE, the Persian Achaemenid Empire controlled an area that on today's map would encompass Pakistan, Egypt, Turkey and everything in between – over 2 million square miles in total.[25] A few centuries later, around 200 BCE, the Han dynasty united an area of similar size in China.[26] Around 100 CE, Rome commanded the entirety of the Mediterranean and most of modern-day Europe, and by 750 CE the Umayyad Caliphate ruled an area stretching from modern-day Iran all the way across to Morocco and into Spain.

One explanation for why the dominant unit of society shifted from city-state to empire focuses on the shift from bronze to iron. Bronze, an alloy of copper and tin, was the material of choice for early human societies when it came to weaponry and armour. Bronze is not only robust, but also relatively straightforward to produce, given the relatively low melting temperature of its constituent metals. The downside of bronze, however, is that it requires access to rare deposits of copper and tin. Iron ore, by contrast, is plentiful, but requires complex furnaces to smelt given the high temperatures that need to be reached. Once societies began to master the process of producing iron, a more abundant supply of weaponry and armour gave them a significant advantage over rivals. The specialist knowledge and investment required for iron foundries made it difficult for the smaller city-states of the ancient world to survive.

Perhaps even more important than the shift from bronze to iron was the emergence of a series of new belief systems that underpinned cooperation at greater scale. While organized religion was a feature of many of the earliest human civilizations, their ancient gods tended to be concerned more with rituals and sacrifices than morality, and with the local community rather than humanity as a whole.[27] Over time, new religions began to emerge that threatened divine retribution for immoral behaviour and expanded divine jurisdiction over all people.[28] One of the earliest examples of this was the Zoroastrianism of the Achaemenid Empire, which was the first religion based around the idea of a single almighty god who would sort humankind into heaven and hell based on their behaviour on earth.[29] Universal moralizing religions such as Zoroastrianism helped diverse communities separated by vast geographical distances to sustain a unified identity.

Secular belief systems also played a role. The teachings of Confucius, roughly coinciding with the birth of Zoroastrianism, embedded ideas of mutual obligation deep into Chinese society where they have remained ever since. For Rome, a political philosophy that afforded rights and protections to conquered peoples played a crucial role in maintaining cohesion across the empire, as did efforts to embed the Latin language and customs in the provinces.[30]

In Rome's case, the empire also used its power to mobilize monumental public works that improved living conditions in its conquered territories, a fact notably captured in the Monty Python film *Life of Brian*. As the comedy troupe plots to overthrow their Roman rulers, the lead conspirator rails: 'Apart from the sanitation, medicine, education, wine, public order, irrigation, roads, a fresh-water system and public health, what have the Romans ever done for us?' While Roman rule could have terrible consequences, not least for slaves, it also led to many advances.

The birth of empires matters for our story as with them came a new type of city on a much greater scale: the imperial capital. While early cities were constrained by the size of their immediate hinterlands, Rome was able to ship in grain from conquered territories such as Egypt to support an unprecedented number of inhabitants for its time. The scale of the Roman Empire made it impossible to govern from one city. That led to the delegation of power to a series of provincial capitals like Alexandria or Corinth, which in turn ruled over a collection of smaller cities in their orbit. Rome nevertheless remained the fulcrum of political, economic and cultural life, and its inhabitants achieved a standard of living unrivalled elsewhere in the empire.

The dominance of the imperial city was not an experience unique to Rome. When the first Han emperor established his capital at Chang'an around 200 BCE, it rapidly grew into the principal city in China, its focal point of power and learning, with a large population of scholar gentry managing the imperial bureaucracy, and an even larger underclass supporting them. As China opened up trade with civilizations to its west, across what would later come to be known as the Silk Roads, the Han capital took on even greater economic importance as a major destination along the trade route, with an extraordinary range of goods flowing in and out of its marketplaces. With the collapse of the Han dynasty, Chang'an went into a phase of decline, but it would return to prominence as the capital of the Tang dynasty in the eighth century, at which point it would become the largest city in the world, with a population similar to Rome's at its zenith.[31]

SPECIALIZATION

In 2015, YouTube star Andy George set out to make a sandwich entirely from scratch. Not merely content with purchasing supplies from his local supermarket and assembling his lunch, he made all the ingredients himself, growing the wheat to bake his own bread, milking a cow to make his own cheese, and raising and slaughtering his own chicken. The process took six months and cost him $1,500.[32]

Today we would not survive long if cut off from the many-layered supply chains that provide us with food, clothing and shelter. The power of the division of labour is one of the foundational lessons of economics. In the opening of *The Wealth of Nations*, Adam Smith describes how workers in a pin factory could only ever produce a small fraction of their potential if they all sought to complete each step of the process on their own, rather than focusing on one specific stage of production. For Smith, specialization was the foundation of material prosperity in society.

The power of specialization rests on two pillars. First is learning effects. The more one repeats a task, the better at that task one gets. Common pitfalls are avoided, confidence builds, and through experimentation smarter ways of doing things emerge. Anyone who has watched an experienced craftsman in action and then tried clumsily to replicate their handiwork will be familiar with this. Second is economies of scale. The wheat required for ten sandwiches is not ten times as costly to produce as the wheat required for one sandwich, as the incremental time and equipment required for each ounce of grain steadily declines.

It was not until the emergence of the city that a complex division of labour became viable. In a small village, there are simply not enough prospective customers for an individual to make a living out of pottery. But as villages grew into cities, an intricate system of specialized activities became possible. Clay tablets from Uruk enumerate a long list of occupations, including potters, gardeners, stonecutters, jewellers, cooks and weavers.[33] As populations gathered together in ever-larger communities, specialization continued to deepen, productivity rose and living standards improved. By the time the Greek historian Herodotus visited Egypt in the fifth century BCE, he could marvel at how 'some physicians are for the eyes, others for the head, others for the teeth, others for the belly, and others for internal disorders'.[34]

Cities also helped propel the division of labour by providing a central marketplace where people could meet to trade goods and services with one another. Centralizing exchange in one place is far more efficient than individuals travelling to numerous locations to procure what they need and sell what they can offer in return. Jesus may have expelled the merchants and moneychangers from the temple, but setting up shop in a street or market where a steady flow of prospective customers is guaranteed makes good business sense.

As well as enabling trade within their boundaries, cities facilitated commerce within and between regions. As far back as the Bronze Age, cities acted as the connecting points of an early globalization centred on the Eastern Mediterranean. Trade first formed along the river systems on which cities were established, with boats providing the means to transport large volumes over long distances. Advances in technology soon allowed these networks to spread further afield. The domestication of the horse and the invention of the wheel made it possible to travel large distances over land, while advances in shipbuilding opened up the possibilities for trade along the Mediterranean. As a result, a long-distance network of trade emerged, connecting cities from the Indus Valley westward to Egypt and up into Greece.[35]

Since then, cities have formed the backbone of a number of other important historical trade routes. The Silk Roads, which stretched from China in the East through India and Persia all the way to the Roman Empire, were linked by an extensive chain of cities. China sold its silk, tea and porcelain; India offered its spices and textiles; Persia contributed gold, silver, copper and iron; and Rome traded with glass, wine and jewels.[36] By the first century CE, Chinese silk was so widespread in Rome that the philosopher Seneca was complaining that the thin, figure-hugging material had become a threat to marital relations.[37] As trade between civilizations along the Silk Roads grew, so did the importance of the cities that facilitated them. Later, during the Middle Ages, the Hanseatic League of cities in Northern Europe connected a trade corridor that spanned from Novgorod in the east to London in the west, facilitating exchange in goods ranging from fur and fish to wool and salt.[38]

The expansion of long-distance trade brought new opportunities to cities that otherwise lacked the magnetic power of an imperial capital. Venice rose to prosperity as an entrepôt facilitating trade between east and west.[39] Later, cities such as Singapore and Hong Kong would grow wealthy by facilitating ocean trade between Asia and the rest of the world.

There are many reasons why cities have always been vital to global trade that go beyond merchants requiring resting places on long-distance routes. Hub-and-spoke distribution systems, with cities as the connecting nodes, reduce the total distance of travel required and the amount of physical infrastructure such as ports and roads needed to facilitate trade. Cities also offer translation, financing and other services that facilitate exchange. These in turn are made possible by the presence of migrant diasporas. In antiquity, as today, cities provide the scale necessary for a minority ethnic community to maintain their traditions and a sense of belonging in a foreign land. Assyrian merchants during the Bronze Age established local settlements in cities throughout Anatolia, in modern-day Turkey.[40] The equivalent today can be seen in the iconic China Towns of London, Sydney, San Francisco and elsewhere. Diasporas have long provided an essential lubricant to global trade by establishing connections with foreign supply chains and providing a support network to visiting merchants.

Trade through most of history was predominately shaped by the natural endowments of different regions. Variations in climate, material deposits and flora and fauna made trade mutually advantageous by allowing economies to focus on producing those goods in which they had the greatest comparative advantage, an insight originally crystallized by the classical economist David Ricardo. In the 1930s, the Swedish economists Eli Heckscher and Bertil Ohlin extended this theory for an industrialized world by also considering variations in the relative amount of labour and capital available in different economies. Those with an abundance of capital relative to labour – rich countries, in other words – would benefit by focusing on the production of capital-intensive goods, such as industrial machinery, importing labour-intensive goods, such as textiles, from abroad.[41]

By the 1970s, it had become apparent that these theories could not adequately explain patterns of trade. In particular, they failed to make sense of the growing amount of trade between countries with very similar endowments. It was the economist Paul Krugman who established the missing link, with a theory for which he would be awarded a Nobel Prize. At its heart lay a newfound appreciation of the role of the city, not the nation, as the building block of the modern global economy.

The United States emerged as a global leader in commercial aviation and Germany as a global leader in machinery not because of the differences in the relative endowments of each country, but because the two countries built specialized clusters for these industries, centred respectively on Seattle and Munich. The origin story of these specialisms is often accidental, but the strength of their advantage grows with time. Learning effects and economies of scale mean it is more productive to have one Boeing and one Siemens, rather than a Boeing and a Siemens in many cities. Globalization opens up a prospective market large enough to justify this level of specialization. Over time, an ecosystem forms around these companies, with aviation suppliers gravitating towards Seattle and machinery specialists towards Munich. The end result is a more efficient distribution of global resources and, in principle, a bigger economic pie for all to share in.

In recent decades, the scale of global trade has expanded significantly. This is partly due to technological advances such as the containerization of shipping, which helped simplify the movement of goods, and the rise of the internet, which helped improve long-distance communication. The other important factor has been a reduction in tariffs, quotas and other trade barriers that isolated many economies. As a result, the share of global output that is traded today is around 30 per cent, far greater than anything seen in history.[42] This is not just about the export of cheap manufactures from Vietnam or financial services from London. Many countries are now heavily reliant on others for the food they need to sustain themselves, not only in developing economies such as Bangladesh or Nigeria, but also in more advanced ones like the United Kingdom, Japan and South Korea.

While the impact of global trade on workers has in recent decades been profoundly uneven – an issue we return to in Chapter 3 – the gains to consumers have been immense, with specialization helping to significantly lower the price of many goods. It is cities that have made this possible, and without them, the complex networks of global supply chains we so rely upon would quickly unravel.

INVENTION

While specialization allows societies to make the most of resources given the current state of technology, it is the power of invention that truly pushes the frontier. Thanks to technological progress, billions of people around the world are now able to achieve a level of material comfort that would have been unimaginable only a few centuries ago. Although there are stark inequalities, and terrible hardship remains the reality for too many, overall rates of malnourishment have plummeted and life expectancy has soared. 200 times as many humans inhabit the earth today as compared with the number at the time of the first city, and they are immeasurably better off materially.[43]

Cities are the cradle of invention. Writing, the plough, the potter's wheel, the sailboat, metallurgy, the calendar and abstract mathematics all emerged within a few centuries of the foundation of the first cities.[44] From the Lyceum of Ancient Athens, to the piazzas of Renaissance Florence, to San Francisco or Bangalore today, human creativity has always flourished in cities. What is true of science and engineering is equally true of philosophy and culture, those things that allow us not only to live better materially, but also to do so with greater expression and meaning.

The power of specialization we described above is one reason cities are wellsprings of new ideas. Aristotle would not have been able to develop his notions of logic, metaphysics and ethics had he been working in fields all day. And Galileo Galilei would not have been able to advance our understanding of gravity and astronomy had he not benefited from the patronage of the Medicis who had grown wealthy on the trade of Florence.

But the intellectual vigour of cities goes well beyond this. While intellectual history tends to be told as a story of great thinkers conjuring up breakthrough ideas in laboratories and bathtubs, the reality is far more subtle, the result of a continual tinkering and recombination of existing ideas.[45] This process thrives in cities, partly because the greater density of population improves the flow of information, and also because the greater diversity of life in cities offers a broader intellectual foundation on which to draw. In the ninth century CE, Baghdad was the capital of the Islamic world, and one of the largest of all cities. It was also a global centre of science and learning thanks to the House of Wisdom, a palatial complex that brought together intellectuals of many ethnicities, languages and faiths to read, write and debate with one another.[46] The foundations of modern mathematics, astronomy, chemistry and medicine can all be traced back to this diverse melting pot.

While Baghdad was in its creative prime, Western Europe was in the depths of its so-called 'Dark Ages' during which scientific and philosophical enquiry received scant attention. There are many explanations for the eruption of ideas that eventually occurred during Europe's Renaissance in the fifteenth century. The fall of Constantinople to the Ottomans undoubtedly played a role, triggering a great migration of Greek scholars to university cities such as Bologna and Padua, bringing with them the canon of ancient philosophy and science that had been lost with the fall of Rome.[47] Also important was the invention of the printing press, which revolutionized access to ideas. Before, knowledge had been locked away in handwritten manuscripts, many of which were kept in monasteries, out of reach of ordinary citizens. With the printing press, books and pamphlets in local languages suddenly became widely available. But while the collapse of the Byzantine Empire and the invention of the printing press helped propel the Renaissance, Europe's intellectual revival was already gaining momentum by the time these things happened. What really laid the foundations for the Renaissance was the return of urbanization following a long hiatus after Rome's collapse, which for the first time in centuries made it possible for people to be regularly exposed to new ideas.

There is good evidence to support the belief that the physical interactions fostered in cities play a critical role in the creative process.

In the 1990s, the psychologist Kevin Dunbar undertook an in-depth study into how breakthroughs emerged across four microbiology laboratories. While time at the microscope was certainly a critical part of the process, discoveries mostly occurred during the lab meetings in which the scientists would discuss hypotheses and challenge one another's findings.[48] Since the time of Ancient Athens, those seeking out answers to humanity's most challenging questions have done so in collaboration with one another.

Technological change has made communication over vast distances easier and cheaper, but it has not diminished the importance of cities as focal points for creativity. During the seventeenth and eighteenth centuries, an informal community of intellectuals formed across Europe that came to be known as the Republic of Letters. Through written correspondence, they would disseminate new ideas and challenge one another's thinking. The network included a range of figures such as Adam Smith, David Hume, Benjamin Franklin, Jean-Jacques Rousseau, Voltaire and many others who together forged the movement of free thinking and enquiry that came to be known as the European Enlightenment.

Far from diminishing the importance of cities as centres of European thought, the Republic of Letters made them even more vital. Gatherings in the salons of Paris, the clubs of Edinburgh and the coffeehouses of London nourished the flow of new ideas.[49] And many of the relationships that underpinned correspondence were forged in cities. As a case in point, it was not until a voyage to London in 1763 that Benjamin Franklin began his correspondence with revolutionary thinkers on the continent.[50] Without the Republic of Letters, Europe is unlikely to have formed the pattern of free thinking and practical enquiry that made possible one of the most consequential phases of technological invention in human history: the industrial revolution.[51]

If we took someone living in Britain in 1000 CE and propelled them forward in a time machine century by century, they might initially be surprised to see how little life changed for future generations. Empires would rise and fall, new lands would be discovered, and ideas about the universe would evolve, but the day-to-day reality of life for most people would look largely the same. But by 1800 they might begin to notice an emptying out of the countryside and an impressive growth in the availability of clothing, pottery and other goods. One more century forward and the world of 1900 would be unrecognizable. In place of small country towns, they would find crowded industrial cities bellowing smoke and churning out vast numbers of seemingly identical products. They in all likelihood would be appalled by the conditions faced by industrial workers, but if they came back at the turn of the next millennium they would find that the rapid expansion in society's productive capacity had dramatically improved the living standards of even the poorest members of British society, with the constant threat of starvation and premature death replaced by a universal expectation of clean water, access to health services, and protection from criminality.

The precise chain of events that led industrialization to occur when, where and how it did is a topic of endless debate among scholars. One explanation emphasizes a shift in agricultural practices and land rights that increased the profits available for landholders to invest in manufactured goods and freed up the necessary supply of labour to produce those goods. Others see a series of technological breakthroughs, including the spinning-jenny and the steam engine, as the root cause, as these innovations made it possible to produce far more goods with the same amount of labour.

What is not in doubt is that, once the process of industrialization was set in motion, it was cities that drove it forward. Early industrial cities like Manchester or Glasgow acted as matchmakers for workers desperate for employment and capitalists eager for profit. At a time before widespread mechanical transportation, factories and their workers needed to be within walking distance of one another. Factories also benefited from being located near to other factories, to stay abreast of the latest inventions and access the engineering expertise that was in such short supply. Cities supported the diffusion and improvement of the techniques and technologies of manufacturing.

As productivity grew, so did the incomes of the middle class, fuelling greater demand for manufactured goods and factory labour. Rising incomes supported investment in more advanced machinery, which in turn contributed to the broadening of the industrial revolution beyond its initial focus on textiles. At the same time, the mechanization of agriculture, and later the introduction of chemical fertilizers, reduced the need for farm labour, drawing even more workers from the farm to the factory. With time, the emergence of the labour movement helped ensure this new working class also benefited from the rising output of the economy through higher wages, which further contributed to rising demand for manufactured goods. The result was a self-sustaining cycle of industrialization, urbanization and rising living standards, the benefits of which have gone far beyond the mere production of more things. Global life expectancy has more than doubled since the onset of industrialization, thanks in part to less malnourishment, but also thanks to a series of advances in medicine and sanitation made possible by the scientific, technological and economic advances that accompanied industrialization.

The industrial revolution profoundly altered the economic geography of the countries in which it first occurred. Manchester, not London, was the centre of industrialization in Britain. In 1780, before it became a manufacturing powerhouse, Manchester had a population of about 30,000, compared to roughly 800,000 in London.[52] Yet the physical geography of the city – with abundant flowing water, plentiful coal deposits and access to the port of Liverpool – made it ideally positioned for industrialization. In 1781, the first steam-powered cotton mill was opened in Manchester; by 1830, the city had a hundred such factories.[53] Its population increased nearly ten-fold over the same time period.[54] And while Manchester was the first place in Britain to be transformed from town to city in a matter of decades as a result of rapid industrialization, it was soon joined by Birmingham, Leeds, Sheffield and others. While London also continued to grow rapidly throughout this period, Britain for the first time since William the Conqueror experienced a meaningful decentralization of economic heft away from the capital and into the regions. In this way, industrialization not only pulled people into cities, but also made possible a flattening of the urban hierarchy that had existed since the birth of the first empires, as the importance of proximity to power waned relative to the importance of coal and industry.

The three forces explored in this chapter have built upon each other through history to make possible the prosperity we enjoy today. Cooperation made possible the density of living and the bonds of connection necessary for specialization. Specialization made it possible for inventors to focus on inventing. New ideas and technologies made it possible for societies to function at ever-greater scale and trade increasingly globally. Far from being passive by-products of the process of human history, cities have been its engine room.

It would be a mistake, however, to imagine this process has been entirely linear. The end of the Bronze Age, around 1200 BCE, was marked by a near-simultaneous collapse in a number of interconnected civilizations around the Eastern Mediterranean. While historians debate the reasons, the underlying cause is likely to have been a period of severe and extended drought precipitated by a sharp increase in northern hemisphere temperatures.[55] As agricultural yields fell, these early civilizations were no longer able to support their growing urban populations, leading to a wave of deurbanization and a reversal in global trade. In the Dark Ages following the fall of Rome, or the periods of civil strife between the dynasties of China, we see further evidence that human society can and does go through major reversals. Many cities suffered devastating reversals during the World Wars, as they did in China during the Cultural Revolution. Forward progress is never guaranteed. And while the expansion of cities in times of growth amplifies the highs, their reversals undermine trade, dampen innovation and weaken cooperation.

Even in good times, the gains from urbanization have been highly uneven. At any point in history, we see some cities thriving and others languishing. And all cities have contained within them deep disparities in wealth and opportunity. The magnitude of these inequalities is not constant through time. In Chapters 3 and 4, we explore the nature of inequality between and within cities, and show what can be done to ensure our urban world is one that brings prosperity to all, not merely the fortunate few.

3

Levelling Up

As far back as the fifth century BCE, the Athenian playwright Euripides wrote that 'The first requisite to happiness is that a man be born in a famous city.'[1] Cities have long divided the fates of those lucky enough to live in thriving metropolises from those who do not. Today, the decline of formerly prosperous cities in the 'Rust Belt' of the United States and other old industrial heartlands is fuelling a politics of resentment and rage. Inhabitants of these communities have found themselves struggling to make ends meet amid downward spirals of joblessness and urban decay. Meanwhile, elites in booming cities like New York, San Francisco, London and Paris have enjoyed soaring incomes. The resulting disillusionment has given rise to Trumpism and Brexit, and fuelled populist resentment against mainstream parties across Europe and the Americas.

In the previous chapter, we explored how the rise of empires was associated with a centralization of power and wealth in imperial capitals. The tectonic shift of industrialization for a time led to a reversal of this process, as newfound geographical advantages turned previously overlooked regions into economic powerhouses. And while regions naturally endowed with waterways or minerals were the first to thrive under industrialization, successive improvements in transportation – from canals to railways to highways – facilitated the spread of industrial cities across Europe, the United States and beyond.

In 1780, before industrialization, London had roughly 15 times the population of Bristol, the next largest city in

Britain, and was far wealthier than any other part of the country.[2] Paris played a similarly dominant economic role in pre-industrial France. Industrialization led to a decline in this relative dominance, as wealth – and population – shifted to fast-growing manufacturing cities. Starting around the 1980s, however, that process went into reverse, with a smaller number of major cities reasserting their dominant role in the economy. To best understand that reversal, we start not in Europe but in the then newly formed republic of the United States, a country whose lack of a dominant imperial capital makes it the perfect place to study the unadulterated forces of economic geography at work.

THE LAND OF OPPORTUNITY

Samuel Slater was not a popular man in his homeland of England. Having spent most of his life working in a cotton mill, Slater had become intimately familiar with the new machinery that was powering the country's booming textile industry. In 1789, he boarded a ship to the United States with a head full of that knowledge, and brought the industrial revolution to Britain's erstwhile colony.

It was in the Northeast of the US that Slater settled, a region whose geography provided the ideal conditions for the development of a manufacturing industry. The combination of abundant access to the rivers needed to power mills and a less favourable climate for farming made manufacturing an appealing proposition.[3] In the South, by contrast, an environment well suited to agriculture – combined with the moral scourge of slavery – kept the region focused on supplying the cotton to feed the mills. The Midwest meanwhile remained isolated and sparsely populated, with an economy heavily dependent on the fur trade. In 1800 a one-way trip from New York to Chicago still took six weeks.[4]

At the close of the eighteenth century, before industrialization had taken off, there was little difference in incomes per capita across the original states of the union.[5] Most of the population continued to live off subsistence agriculture, with fewer than 10 per cent inhabiting the fledgling nation's early cities.[6] By the middle of the nineteenth century, however, the early growth of manufacturing in the Northeast meant inhabitants of that region had become twice as rich and twice as urbanized as those in other states, with factories in cities like New York, Philadelphia and Baltimore drawing inhabitants from their hinterlands and beyond.[7]

With time, the advantage of the Northeast began to dissipate. The development of the canal system and later the railway over the course of the nineteenth century reduced the time and cost of transporting goods to and from the Midwest, while that region's ample supply of iron ore and coal gave it a distinct advantage in heavy industries such as steel and machinery that were playing an increasingly important role in industrialization.[8] As manufacturing spread eastward from the urban centres of the Northeast, an integrated 'manufacturing belt' emerged, stretching from the Atlantic coast to Minnesota and Missouri.

For the Midwest, the shift to a manufacturing-based economy drove a step change in urbanization and set in motion a convergence in incomes with the Northeast. By the time of the 'Gilded Age' in the closing decades of the nineteenth century, the Midwest was home to some of the richest figures in the nation, including Cleveland's John D. Rockefeller, founder of Standard Oil. The rise of the Midwest continued well into the twentieth century, and by the 1950s income per person in the region had largely converged with the Northeast.[9]

The experience of St Louis is illustrative of the coming of age of the Midwest. St Louis was just a small town in 1800, but by 1900 it had grown to become the fourth largest city in the nation thanks to its role as a major centre for the manufacture of hardware and furniture, among other things.[10,11] Its rising status was signalled by its role in 1904 as host to both the World Fair and the Summer Olympics. St Louis continued to prosper over the course of the twentieth century, with income per capita reaching 90 per cent of that of New York by the end of the 1970s.[12] The city was not without its problems – starting in the 1950s, rapid suburbanization led to the hollowing out and impoverishment of St Louis' urban core, an issue we return to in the next chapter. But the wider metropolitan area continued to thrive until late into the twentieth century.

In contrast to the Midwest, for the South the nineteenth century set off a period of relative economic decline due to the slow adoption of manufacturing and the collapse of its slave-dependent agricultural economy following the civil war. It was not until after the Great Depression that incomes in the region began to reverse their decline relative to the rest of the country.[13] The development of the national highway system, the declining cost of trucking, and the emergence of containerized shipping supported the dispersion of the US manufacturing base into the South through the middle decades of the twentieth century, bringing with it an accelerated catch-up in urbanization rates. Lower wages and weaker legal protections for labour in Southern states also helped, with manufacturers such as Ford shifting production to cities like Louisville in a bid to lower costs. By the 1980s, the spread of manufacturing had brought incomes across the various regions of the US largely in line with each other.[14]

Rural areas also found themselves profoundly altered by the process of industrialization. Through the nineteenth and into the early twentieth centuries, soaring demand for labour to feed urban factories led to a growing disparity in incomes between the countryside and the city. Following the development of the highway system, however, rural towns with their cheap and plentiful land became increasingly attractive locations for factories, which led to a significant drop in the urban wage premium in the post-war decades.[15] In parallel, the industrialization of agriculture saw small-scale subsistence farming replaced by large-scale commercial enterprises that were highly productive but required far fewer people. Of those who stayed behind in rural communities, some found work on commercial farms, but many found employment in new local factories, and an even greater number began to provide the services – from healthcare and education to retail and hospitality – that come with disposable income. Altogether, the result was a transformation of the nature of rural communities, as populations shifted from farmsteads to local towns containing tens of thousands of residents.

Since the 1980s, however, economic activity has concentrated in a small number of major cities such as New York, San Francisco and Chicago, while many formerly thriving cities and towns have been left behind. Cities like Detroit, Cleveland and Milwaukee, once among the richest in the country,[16] have entered seemingly intractable cycles of urban decay, with high rates of poverty, deteriorating public finances, decaying infrastructure and widespread crime. Income per capita in St Louis has fallen from 90 per cent of the level in New York in the late 1970s back down to 67 per cent.[17] The urban-rural income gap has also surged to historic heights, as numerous rural centres have found themselves wrestling with joblessness and stagnating incomes.[18]

The decline in the tyranny of distance through the nineteenth and twentieth centuries helped to spread economic opportunity across the United States, and turned the economy into a network of many thriving cities and towns. Britain, France and other industrialized economies enjoyed a similar experience. Why then have we witnessed such a dramatic divergence over the past four decades? The answer lies in the interplay between three structural shifts in the economy: deindustrialization, superstar dynamics and declining mobility. Together these shifts make sense of why place has reasserted itself as a major factor in determining the opportunities one receives in life, and what must be done in order to create a more level playing field.

DEINDUSTRIALIZATION

In the US, the share of workers in manufacturing jobs has fallen from around 25 per cent in the 1950s to around 10 per cent today,[19] an experience echoed in Britain, France and other rich countries. Cities like St Louis or Newcastle, where inhabitants were heavily dependent on the jobs created by local manufacturing industries, have struggled to adapt. In her moving book *Janesville*, journalist Amy Goldstein vividly demonstrates the cascading impact of the closure of a manufacturing site on the local economy, as suppliers close shop, laid-off workers struggle to retrain, and local retailers and service providers watch their sales dry up.[20]

Globalization bears much of the blame for the disappearance of manufacturing jobs, with production moving to locations such as China and Mexico to take advantage of lower labour costs and cheaper land. But the transition to less labour-intensive forms of production has also been important. Since the early 1990s, the number of workers it takes to produce a car in the US has fallen by roughly 25 per cent thanks to greater use of labour-saving machinery.[21]

The auto industry is just one of many examples of this trend. Economists break down economic activity into an amount contributed by labour, which is equal to the wages of employees, and an amount contributed by capital, which is everything else. Since the start of the 1980s, the contribution of labour in the manufacturing sector in the US has fallen from over 70 per cent to under 50 per cent, as machines have increasingly replaced workers on the factory floor.[22] As industrial robots continue to become more sophisticated and widespread, this trend is likely to accelerate further.

Manufacturing is not the only sector that has been heavily affected by offshoring and automation. Between 1940 and 1980, office administration jobs, including clerical and secretarial work, doubled from roughly 10 per cent of all employment in the United States to nearly 20 per cent, as increasing numbers of staff were required to manage the growing scale and complexity of ever-larger enterprises.[23] Since then, digitization of record keeping and correspondence, combined with the offshoring of operations to India, the Philippines and other low-wage economies, has reduced the share of these jobs to below 15 per cent of total employment in the US. This figure will continue to decline in the decades ahead as algorithms replace a growing share of activities such as payroll management and accounts payable. The loss of such jobs has added to the challenges faced by many smaller cities, as regional headquarters and back-office processing centres have disappeared.

Understanding where the jobs have gone is essential for understanding the growing divergence among flourishing and languishing places in the United States and other rich countries. The changing job market is a tale of two halves. On one hand, demand for high-skill knowledge professionals has soared. Finance jobs are among those that have experienced the greatest growth since the 1980s, as the increased size and complexity of financial markets has driven up demand for bankers, traders, fund managers and financial engineers. While the global financial crisis took much of the heat – and shine – out of the sector, it did little to structurally reduce the employment share of such activities. Other white-collar professionals, including lawyers, marketers, business consultants and executives, have also seen their role in the overall occupation mix expand significantly. More recently, rapid growth in high-tech sectors like digital services and medical technology has led to a surge in demand for science and technology specialists, including software engineers, data architects and applied research scientists.

Many of those who lost their jobs in manufacturing or administrative roles in recent decades have not been able to move into these high-skill, high-pay jobs. Nor have their children. Instead, they have shifted down the economic ladder into low-skill, low-pay service jobs, the only thing available for many of the displaced labour force. Booming demand from busy, highly paid professionals for childcare, online deliveries and other conveniences has driven up the need for these types of services. And while the ageing population of rich countries has created more openings for doctors and other highly paid healthcare professionals, the majority of jobs created have been in low-paid care roles in hospitals and nursing homes.

Through much of the twentieth century, the dispersion of manufacturing and administrative jobs played a central role in widening opportunities and raising incomes. These middle-skill jobs provided those without a college education with a decent income and a shot at upward economic mobility. With their disappearance, a polarized job market has emerged with profound implications not only for individuals, but also for many formerly prosperous cities and towns.

SUPERSTAR DYNAMICS

In 1890, the economist Alfred Marshall coined the term 'economies of agglomeration' to describe the benefits to companies from locating near to one another.[24] Marshall identified that access to labour with the right skills, proximity to suppliers, and the informal transfer of knowledge across companies could explain the geographic clustering of industries such as textiles in Manchester or metalworks in Sheffield. In subsequent knowledge-based economies, these forces have become even more significant, giving rise to a number of 'superstar' cities, including San Francisco, New York, London, Paris and Shanghai. These cities have captured the lion's share of economic growth in recent decades.

The first reason for the emergence of these superstars is talent clustering. It was relatively straightforward to train up a worker to perform the manual tasks that once characterized most manufacturing activity, compared to the biomedical scientists, software developers and financial engineers that are critical to today's economy. While we might like to believe that the internet now allows anyone to become an instant expert in anything, the reality is that many of the high-skill occupations that drive economic activity today rely on years of accumulated knowledge through both formal education and apprenticeship on the job. Many of these occupations have become highly specialized – a biomedical firm is not just looking for any available worker, but rather someone with expertise in the specific field in which they operate. The result is that companies in many fast-growing industries want to be located in cities where they feel confident they can access the types of skills they require. This further reinforces the benefit to workers from locating in such areas, in a virtuous cycle of increasing talent density.

A second reason for the increasing strength of agglomeration effects is the way in which knowledge workers in particular become more productive when located near to one another. Information may be more available than ever, but the evidence suggests that it still tends to flow largely through in-person networks, spurred on by casual encounters within and outside the workplace. Inventors filing for patents are twice as likely to cite patents from their own city than from other places, even after excluding those generated within the same company.[25] The physical proximity of people in cities is particularly important when it comes to drawing inspiration and knowledge from diverse fields. Economists Enrico Berkes and Ruben Gaetani analysed the references cited in patent filings in the US between 2002 and 2014, and found that densely populated areas in the US were far more likely to produce patents based on 'unconventional combinations of prior knowledge'.[26] Given the importance of intellectual property in many of the industries driving economic growth in rich countries today, it is unsurprising companies are drawn towards those superstar cities where they are more likely to be able to absorb cutting-edge ideas.

A third reason for the growth of superstar cities in recent decades is the growth of superstar firms. The benefits of scale for firms are stronger than ever in today's knowledge economy. US companies with revenues in excess of $10 billion generate a return on equity that is nearly four times the rate generated by firms with revenues less than $1 billion, up from 1.2 times in the 1980s.[27] Part of this is the rapid growth in industries like social media that exhibit 'network effects', where the value of a service to users increases with the number of other users, creating a natural tendency towards a single dominant player. However, the widening differential in company performance can also be seen in many industries that do not experience these kinds of effects, suggesting there is more to the story. In a world of greater automation, delivering products or services typically involves more upfront investment, but lower per-unit cost once those initial investments have been made. That dynamic favours larger firms who can spread these costs over a bigger revenue base. The result has been a phase of increased consolidation across many US industries, from banking to airlines to consumer goods, a process helped along by a shift in anti-trust regulation in the 1980s that made it easier for firms to acquire one another.

Experts disagree on whether increased market concentration has been good or bad for consumers. If weaker businesses are hoovered up by the most efficient companies, it can lead to lower prices. But less competition can also increase the ability of companies to inflate their prices and reduce the choice of products available. And what might be beneficial in the short term can undermine dynamism and productivity in the medium term, as monopolistic firms lose the incentive to keep innovating.[28] There is already worrying evidence of this occurring – in 2011 spending by Apple and Google on patent litigation and purchasing overtook spending on research and development.[29]

The increasing consolidation of industries and domination of a growing number of sectors by superstar firms has bolstered the rise of superstar cities. In 1980, St Louis was home to 23 Fortune 500 companies. Today, that figure is down to eight.[30] While some of these businesses simply failed, others were acquired in the waves of industry consolidation that have swept across the US. Local aerospace firm McDonnell Douglas was acquired by Boeing in 1997. Ralston Purina was purchased by Nestlé in 2001. Anheuser-Busch was purchased by InBev in 2008. As industries became more concentrated, headquarters were consolidated into a small number of leading cities, leaving places like St Louis bereft of jobs.

On the whole, cities that were already large have been far more likely to enjoy robust growth and rising incomes in recent decades. There are certainly some exceptions to this pattern. Austin was the 67th largest metropolitan area in the US in 1980, but has since experienced impressive growth in both size and income level; it is now the 28th most populous metropolitan area in the US, with a GDP per capita that has increased 40 per cent, adjusting for inflation, over the past two decades.[31] Detroit, meanwhile, was still the fifth largest metropolitan area in 1980, but has struggled in recent decades; it is now the 14th largest metropolitan area, with an inflation-adjusted GDP per capita that has increased only 8 per cent in two decades, well below the national average. These examples are interesting exceptions, which offer valuable lessons around the potential for reversing the fate of struggling cities, as well as the importance of avoiding complacency. But they are not the norm.

The fact that, in general, large and successful cities have tended to get even larger and more successful suggests that firms and workers experience spillover benefits from being located nearby not just to others within their industry, but also to big cities more generally. It is not a coincidence that thriving cities today are often leaders across multiple sectors. San Francisco is famous as a major hub for digital technology, but it is also a leading city for biomedical research and a major hub for financial and professional services on the West Coast. New York is not just home to Wall Street, but is also the focal point of the US fashion industry. London and Paris host the headquarters of most major companies in Britain and France respectively.

Multiple factors contribute to the positive spillovers firms and workers enjoy from locating in major cities. Company value chains often cross industry boundaries. Proximity to financial and professional services, for example, is an important drawcard for many large cities. The presence of top universities is another factor. The same universities that produce investment bankers also produce lawyers, data scientists and medics. Lastly, university-educated professionals typically choose to live in places where other university-educated professionals live. Partly this is about humans' natural tendency to gravitate towards people with similar outlooks to themselves, what sociologists call 'homophily'. Increasingly, it is also about romance. In the 1970s, roughly 50 per cent of individuals with a bachelor's degree were married to a spouse who also held a bachelor's degree. Today, the figure is closer to 70 per cent,[32] with highly educated workers increasingly seeking out partners of a similar education level. And as dual-career households have become the norm, these couples need to find places to live that offer good opportunities in both of their lines of work. If a lawyer marries an advertising executive, or a scientist partners with a management consultant, they are likely to look to large cities that have a critical mass in both industries, further adding to the gravitational pull of these places.

With the growing number of skilled workers in superstar cities has come increased jobs for low-skilled workers too. To understand why, it is helpful to differentiate between the 'tradeable' and 'non-tradeable' sectors of the economy. The tradeable sector produces things that can be consumed away from where they are originally produced. Manufactured goods, digital products and most financial services fall into this category. The non-tradeable sector, by contrast, encompasses services that are intrinsically anchored in place, such as hospitality, construction and most healthcare. When a city brings in new jobs in the tradeable sector, these workers also consume the services of the non-tradeable sector, creating a multiplier effect for the local economy. While some of these are high-skill jobs – for example, doctors, lawyers, architects and chefs – the majority are low-skill roles like waiters, cleaners, childcare workers and construction labourers. The economist Enrico Moretti has studied these multipliers, and found that the creation of one new tradeable sector job in an area generates demand for 1.6 new jobs in the non-tradeable sector, as these workers go to restaurants, get haircuts, visit the doctor and build new houses.[33] However, this multiplier effect is far stronger for jobs in high-skill tradeable sectors – say at a company like Apple – where one new job in an area can generate as many as five new non-tradeable jobs in its wake.[34] This is not just because high-skill workers tend to have greater incomes, but also because they are often time poor and more likely to eat out and spend on services such as household cleaning.

The growth of superstar cities has been further reinforced by their role as magnets for foreign migrants, both skilled and unskilled. While London contains 13 per cent of the UK's population, it is home to 35 per cent of those living in the UK who were born abroad.[35] As noted in Chapter 2, cities have long played an essential role in facilitating global mobility by providing the scale necessary for meaningful diasporas to form. And the larger the city, the greater the variety of diasporas that can reach threshold scale, which is why big cities are particularly effective in drawing in migrants. While many immigrants enter low-skill service occupations, they are also disproportionately likely to start businesses that create jobs. In the US, immigrants account for 14 per cent of the population but 25 per cent of new business creation.[36] The share is even higher in technology clusters, with 63 per cent of start-ups in the San Francisco Bay Area founded or co-founded by immigrants.[37] The increase in foreign migration flows in recent decades has therefore further bolstered the gravitational pull of major cities and contributed to their outperformance.

DECLINING MOBILITY

In an economy with a fully mobile labour force, large differences in living standards across cities should in theory dissipate over time. Workers will relocate from places where wages are low to places where wages are high. As they do so, they will decrease the supply of labour in the places they are leaving and increase the supply of labour in places they are going, which should gradually bring wages back in line. Yet it is getting harder and harder to move, as reflected in declining rates of mobility within countries. In 1980, one in every 35 US citizens moved across state borders each year, a figure that by 2020 had halved to one in every 70.[38] If low-skill workers were to migrate out of declining cities and towns, it would help to break the vicious cycles of high unemployment, depressed wages and deteriorating living standards experienced in these places. Why have we not seen this happen?

Beyond the obvious and profound ties of personal history, family and friendships that connect individuals to communities, there are structural forces at play that have lowered the propensity for workers to relocate in search of better opportunities. One is that the population in rich countries today is far older than it has ever been before. As workers age, they tend to relocate with lower frequency,[39] as they are more likely to have established a family, be caring for elderly parents, or own a property that would need to be sold. The transition to a predominately dual-income family structure has also worked to reduce mobility. Relocating now typically involves two job searches rather than one, each bringing with it a high degree of uncertainty. And while one half of a couple may be struggling to find work, the other half may be quite happy where they are.

The reasons for the steep decline in mobility run even deeper, and relate to the emergence of the polarized job market described earlier. Historically, less educated workers had a relatively higher propensity to relocate, as good middle-class manufacturing or administrative jobs created many opportunities for a better life in more prosperous areas. Now, less educated workers looking to relocate to superstar cities are typically only able to secure low-paid service jobs. And while wages for these jobs tend to be higher in thriving cities, the difference is often more than offset by the increased cost of living. A low-skill worker moving from Cleveland to San Francisco would on average enjoy an after-tax increase of roughly 30 per cent, but a *decline* in their spending power of roughly 10 per cent, after adjusting for the increased cost of living, leaving them worse off.[40]

By far the biggest factor behind this disconnect is rising house prices. In Britain, for example, the ratio of house prices to average incomes has risen from three times in the mid-1990s to nearly seven times today.[41] In London, this ratio has gone from roughly three and a half times to 11 times, while in the North of England it has only risen from three times to four times. All across the rich world, house prices in major cities have soared in recent decades. We explore the causes of this rise, and what can be done about it, in the next chapter. For now, we simply note that the increasing unaffordability of housing in sought-after places has acted as an important brake on low-skill households moving to areas that offer greater economic opportunity.

In theory, education should allow less educated workers to transition into knowledge professions. Education is one of the most powerful forces for upward economic mobility, even more so as the earnings gap has widened between high-skill and low-skill workers. In practice, it is not that easy. For one thing, as the workforce ages, the economics of reskilling mid-career start to weaken, as there are simply fewer working years remaining to make a return on the investment of time and money. Even more problematic has been the increasing unaffordability of education. In the US, the cost of a four-year college degree adjusted for inflation has nearly tripled since 1980, and while alternative paths to reskilling, such as micro-credentials, have become more widely available, many employers have been slow to accept them.[42] Add to this the fact that fully 40 per cent of recent US university graduates end up in a job where a degree was not necessary and it becomes easy to see how many from low-income backgrounds may lack confidence in higher education providing a path to a better life.[43] For young people in left-behind places, peer effects compound this story. When very few of those you grew up with are choosing to go to university, it takes a great deal of courage to break away from the crowd, particularly if you are uncertain of your prospects for success.

Those that do end up attending university often choose to leave upon graduation, drawn to the attractive career and life prospects offered by the hubs of the knowledge economy. Washington University in St Louis and Warwick University in

Coventry are both excellent academic institutions, but neither city has managed to hold on to a sufficient quantity of its top graduates to build deep professional clusters. Good universities are a necessary but not sufficient condition for a successful city in today's economy.

RESTORING BALANCE

It may be tempting to resign ourselves to the prospect of an ever-widening disparity in outcomes between those lucky enough to be in thriving cities and those who are not. Yet there are reasons for optimism. In the 1970s, Seattle had entered a cycle of seemingly unstoppable decline. A large billboard on the way to the airport appeared reading: 'Will the last person leaving Seattle turn out the lights.'[44] Since then the city has enjoyed a renaissance, with a combination of robust population growth and one of the highest metropolitan incomes per capita in the United States. Microsoft, Amazon and Starbucks all call Seattle home, and while Boeing has since moved its corporate headquarters first to Chicago and now near to the Pentagon, the city remains the primary hub for the commercial aviation industry in the US. We will return later to the reasons for this dramatic turnaround.

Another reason for hope is that there are economies where living standards are far less varied across cities. Figure 2 compares the cities of the US, Britain, France, Germany and Japan. Each bubble represents a city, and the size of the bubbles represent that city's share of total urban population in their country. The bubbles have been ordered vertically based on the average income per person of each city relative to the average income for all urban dwellers in their country. A few salient points stand out. The first is that the urban population of the United States is spread across many cities, with no single city dominating. This is a result of both a large population and a large physical landmass. At the same time, however, the cities of the US are highly varied when it comes to per capita incomes. In the San Francisco Bay Area, the top bubble on the chart, the average inhabitant is twice

FIGURE 2: Income dispersion across cities varies from country to country

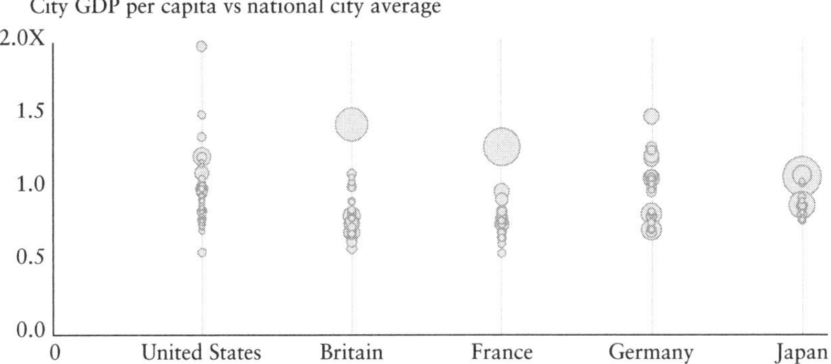

Note: Bubble size represents share of total urban population; Excludes cities with <0.5% of total urban population.
Source: Authors' calculations based on Oxford Economics data.

as rich as the average city dweller in the US. By contrast, the inhabitants of Riverside, the bottom bubble on the chart, are just half as rich as the overall urban average. Britain and France differ in their economic geography from the US, but are similar to one another. Both have a dominant capital city that is much larger in population than any other city in the country, as well as much richer. In this figure, the large bubbles at the top for Britain and France represent London and Paris respectively.

In Germany, the picture is more nuanced. West Germany has multiple thriving urban centres – notably Munich, Frankfurt, Stuttgart, Hamburg and Düsseldorf – none of which is dominant. But it is also home to the Ruhr region, the original heartland of German industrialization, which has since the 1960s been stuck in a persistent cycle of decay as it has struggled to adapt to the decline of Germany's coal and steel industries. And while the major cities of East Germany – particularly Dresden, Leipzig and Erfurt – have been steadily converging with the West since reunification, they remain meaningfully poorer than cities like Munich. Berlin, despite its status as Germany's capital city, also remains much poorer than the cities of the West as a result of its divided legacy, though it too has enjoyed significant convergence in recent years.

If Germany offers reassurance that convergence is possible in a modern economy, Japan demonstrates what can be achieved through a sustained policy effort across multiple decades. In the reconstruction period following the Second World War, a tripolar economy emerged in which Tokyo, Osaka and Nagoya became increasingly dominant economically.[45] Starting in the 1960s, the government of Japan adopted an explicit policy of decentralizing production and ensuring a reasonably consistent standard of living across cities, partly through a combination of tax breaks and subsidies to influence the location choices of local producers, but also through heavy investment in world-leading transportation infrastructure – most notably its high-speed rail network – to dampen the incentive for businesses to all gravitate towards the largest cities.[46] Since then, Japan has experienced many of the same pressures towards agglomeration that we explored earlier in this chapter. This has led to an increased clustering of population into Greater Tokyo in particular, which today accounts for 45 per cent of the country's urban population.[47] Unlike the situation with London or Paris, however, average incomes in Tokyo have not grown meaningfully higher than in other Japanese cities thanks to continued commitment by the national government to balanced regional development. The fact that average incomes are broadly similar across Japan's cities can be seen in the tightly clustered pattern in Figure 2.

What then should be done to restore prosperity and opportunity to the declining industrial cities of the rich world? There is unfortunately no silver bullet. While there is good evidence that high-speed rail has supported a balancing of incomes across cities in Japan, the experience of France suggests that caution is needed in drawing any global lessons. High-speed rail between Paris and Lyon in fact facilitated a further concentration of economic activity in the capital, with a number of companies relocating their headquarters to Paris as a result.[48] A similar danger faces projects like HS2 in Britain, under which high-speed rail services will connect

London with the Midlands and the North. Indeed, the New Economics Foundation, a think-tank, has predicted that 40 per cent of the economic benefits of the proposed rail network will accrue to the capital.[49] Rather than focusing on better connecting the regions to the capital, an alternative strategy would be to use high-speed rail to connect regional cities with each other in order to stimulate agglomeration benefits outside of London. For instance, the 900-square-mile triangle encompassing the four major cities of the North – Manchester, Liverpool, Sheffield and Leeds – is home to roughly 8.5 million people, four leading universities and an Atlantic port. It also benefits from proximity to the attractive Peak District and the Yorkshire Dales. A similar argument could be made for connecting Birmingham with the other major cities of the Midlands, including Leicester, Nottingham and Coventry, and while the distances in the United States are greater, it is not too big a stretch of the imagination to envisage a similar configuration for the cities of the Midwest.

Anchor institutions are a second important pillar for levelling up struggling cities. The decision by Bill Gates and Paul Allen to relocate Microsoft from Albuquerque in New Mexico – where its sole customer was based – to Seattle – where the co-founders were originally from – transformed the fate of the city.[50] By the time Jeff Bezos chose Seattle for Amazon's original headquarters, the presence of Microsoft had already transformed the city into a technology hub. The challenge with anchor institutions, of course, is how to attract them. Struggling cities cannot simply wait for the stroke of good luck that occurred for Seattle. We noted earlier in the chapter that a high-performing local university is not a guarantee of success, but it can certainly help. The presence of Dell has been pivotal in Austin's transformation into a thriving technology centre in recent decades. Born in Houston, Michael Dell moved to Austin to study at the University of Texas, one of the top academic institutions in the South, and stayed on in the city after dropping out of college to focus on his business. Similarly, the thriving life sciences clusters in San Diego – and Cambridge and Oxford in Britain – are strongly connected with the presence of leading bio-science departments at their local universities. Boston offers another instructive example – the presence of top-performing universities including Harvard and MIT has been crucial for the city's ability to transform itself in recent decades from a struggling manufacturing hub to a focal point for professional services, finance and technology.[51] But this is not a process that can be completed overnight. As the former Senator and US Secretary of Labor Daniel Moynihan once quipped, 'If you want to build a great city, create a great university and wait 200 years.'[52]

The liveability of a city also plays an important role in shaping its economic success in today's economy. A robust job market is undoubtedly a major factor in shaping where workers choose to live, but it is not the only factor. The urbanist Richard Florida has studied in detail how cities, by creating an attractive lifestyle for their inhabitants, can draw in the high-skill talent that is essential to so many of the industries that drive economic growth today, setting in motion the flywheel of agglomeration effects.[53] For instance, the edginess of Berlin has helped turn the city into a start-up hub, while a thriving music scene and nightlife has long helped bolster the appeal of Austin. But being trendy in most cases will not be enough – New Orleans, for instance, is a city with rich history, a thriving cultural scene, excellent cuisine and interesting architecture, but it has struggled to maintain a robust economy. Other aspects of quality of life are equally essential. The city of Fukuoka in Japan, for example, has established a reputation as a highly desirable place to live for high-skill workers on the basis of its excellent record on safety, pollution, healthcare and childcare, combined with its lower cost of housing.[54]

A final factor to consider in levelling up left-behind places is governance and public financing. Cities like Leipzig in Germany have benefited from the combination of devolved decision making and federal redistribution of tax revenue to support basic services like public transport and education. By setting its tax rates lower than many cities in the West, Leipzig has been able to incentivize companies such as BMW to set up local operations.[55] Leipzig has also benefited significantly from the Innovation and Structural Change Fund set up by the federal government in 2000 to support convergence between East and West Germany by funding innovative start-ups and scientific research in poorer regions of the country.[56] Fukuoka has the highest start-up rate in Japan thanks to an entrepreneur-friendly business environment in which the local government subsidizes the cost of company registration and provides grants and legal advice to those wishing to build a business.[57]

None of these remedies is likely to be sufficient on its own. Ultimately, what is needed to reverse the fate of struggling cities is a holistic, coordinated effort over multiple decades that engages all levels of government as well as the business and wider community. This is possible, and it is also necessary if we wish to heal the wounds of resentment that plague many rich countries today. Yet even this would only be a job half done. A city needs not only to be prosperous on the whole, but also to provide opportunities for all its inhabitants, rather than just a privileged few. That is the challenge we turn to in Chapter 4.

4

Divided Cities

In *The Republic*, Plato wrote that every city 'is in fact divided into two: one the city of the poor, the other of the rich'.[1] Cities have changed dramatically since Plato's time, yet the observation remains as true as ever. Statistics pointing to the prosperity of places like London or New York hide the fact that these cities are not only home to the immensely wealthy but also places of poverty and hardship for many. And while the adversity faced in the poorest areas of struggling cities is especially acute, these cities also contain neighbourhoods like the East Side of Milwaukee or Altrincham in Manchester that are among the richest in their respective countries.

While cities have always been divided, the middle-class jobs on offer to their inhabitants in decades past made them ladders out of poverty for many. By the middle decades of the twentieth century, jobs in manufacturing, clerical and other semi-skilled occupations generally offered respectable pay to workers who lacked a university education but were willing to move from country to city in pursuit of a better life.[2] Using occupational rankings as a proxy, researchers have found that a boy born in the US in 1940 enjoyed roughly double the rate of upward socioeconomic mobility compared to one born a century earlier.[3] But since then, upward mobility rates have fallen back down.[4] The changing occupational structure of the economy has left many trapped in troubled neighbourhoods through lack of access to decent schools, affordable housing and effective public transport.

Too many go through life never able to contribute what they might to society due to the neighbourhoods in which they find themselves. But cities can still play a vital role in helping everyone to reach their fullest potential in life. To understand how, we first need to understand the ways in which the structure of privilege and disadvantage has shifted within cities since industrialization.

INDUSTRIAL HORRORS

Manufacturing jobs are today viewed with a good deal of nostalgia, as respectable, well-paying work for those without higher education. Yet this was certainly not always the case. Between what William Blake described as the 'dark satanic mills' of early industrialization and the solidly middle-class manufacturing jobs of the middle twentieth century lies more than a century of struggle for fair wages and decent working conditions. For the poor and uneducated drawn into industrial cities at the opening of the nineteenth century, life was a struggle.

Alexis de Tocqueville, the intrepid traveller, was an early witness to the nature of life in Manchester during the first decades of industrialization. Travelling to the city in 1835, he observed 'the crunching wheels of machinery, the shriek of steam from boilers, the regular beat of the looms, the heavy rumble of carts', noises that were 'not at all the ordinary sounds one hears in great cities'.[5] Particularly striking were the living conditions of the city's poor. In his writing, Tocqueville described the dark, crowded and unsanitary workers' quarters surrounding the city's factories as 'the last refuge a man might find between poverty and death'. Particularly moving is his discussion of underground cellars into which 'twelve to fifteen human beings are crowded', which he characterized as 'damp, repulsive holes'. By one estimate, such cellar dwellings contained 12 per cent of the city's population in 1835.[6]

Another notable traveller to Manchester around the same time was Friedrich Engels, who in 1842 went to work at the headquarters of his family's textile company. In his 1845 book *The Condition of the Working Classes in England*, Engels described how the working poor of Manchester lived 'in wretched, damp, filthy cottages' in which 'no cleanliness, no convenience, and consequently no comfortable family life is possible'.[7] Originally sent by his family in an attempt to cure him of his radical political views, his experience had the opposite effect. Engels' experience of Manchester forged him into one of the most ardent critics of early capitalism. The financial support he subsequently offered to a radical political economist – Karl Marx – would prove pivotal in the rise of both communism and the labour movement over subsequent decades.

It is hard now to imagine just how horrid life was in cities during the early decades of industrialization. In 1830s Manchester, the typical working week for men, women and children was close to 70 hours,[8] during which time they would work amid deafening noise and oppressive humidity. The combination of a lack of clean drinking water and overcrowded housing meant diseases like cholera, typhus and influenza were rife.

As a pioneering industrial city, Manchester experienced stark inequalities, but a similar story can be found across other industrializing metropolises of that era. Charles Dickens famously brought the plight of London's working poor to life in novels like *David Copperfield* and *Oliver Twist*. In the US, the inhumane working conditions and unsanitary practices of Chicago's meatpacking district, known as the Union Stock Yards, were brought to the public's awareness by Upton Sinclair's fictionalized 1906 account *The Jungle*. Of Chicago's working poor, Sinclair wrote: 'They are penned up in filthy houses and left to rot and stew in misery, and the conditions of their life make them ill faster than all the doctors in the world could heal them.'[9]

As industrialization led to the mass migration of the poor from the countryside, it profoundly altered the structure of cities. Marketplaces and merchant's houses were replaced by factories and warehouses, surrounded by slum-like housing for workers who lacked any mode of transportation. As the merchant class evolved into an increasingly prosperous bourgeoisie, those with the financial means escaped the pollution and visible destitution of the inner ring of the city, aided by the development of the streetcar and intra-city rail systems.[10]

In 1922, the sociologist Ernest Burgess developed his Concentric Zone Model to describe the typical shape of the city emerging from industrialization.[11] In it, a central business district is surrounded by a factory zone, around which the working class lives in low-quality housing, followed by a ring of middle-income neighbourhoods, and lastly a commuter zone for the wealthy. Based loosely on the economic geography of Chicago, it is a highly stylized model, and exceptions are not difficult to find. Paris kept its factory districts well away from the grand boulevards constructed by Baron Haussmann in the middle of the nineteenth century. London concentrated its industrial activity and polluting factories in the East End, as the prevailing wind comes from the West, keeping the cleaner air of the West End reserved for the city's elite, and maintaining a separate financial district in the historic City of London. Nevertheless, Burgess's model is a helpful guide to understanding the broad pattern taking shape in the rapidly industrializing cities around the world.

While the flight of the well-off from urban centres stretches back into the early stages of industrialization, it took another invention – the automobile – to accelerate this centrifugal movement. At the 1939 New York World Fair, one particular exhibit called Futurama stole the show. The acre-sized animated model, sponsored by General Motors, portrayed a city of the future in which interconnected multi-lane highways with a steady flow of traffic deposited workers from their spacious homes in the suburbs to their offices and back again. Futuristic at the time, the model would provide an accurate prediction of the evolution of US cities over the coming decades, as they were redesigned for the automobile age.

FLIGHT TO SUBURBIA

Despite being central to the rise of Detroit, Henry Ford was no great fan of cities. In one opinion piece in Ford's self-published weekly newspaper, the entrepreneur described the modern city as 'a pestiferous growth', 'the most unlovely sight this planet can offer', a 'suppression of all that is sweet in its natural environment', and 'concentrating within its limits the essence of all that is wrong, artificial, wayward, and unjust in our social life'.[12] Ford was unequivocal in his belief as to the solution: 'We shall solve the City Problem by leaving the city.'

By pioneering the mass production of automobiles, Ford played a key role in the abandonment of urban centres over the second half of the twentieth century. Living outside the polluted and crowded urban centre was once a luxury few could afford. The invention of the car brought that life within reach of the average working family. From 1945 to 1955, the number of car registrations in the US doubled from 26 to 52 million, a per capita increase of 62 per cent. By 1975, the figure would rise by another 65 per cent.[13] The construction of housing further from urban centres was given impetus by the Federal Housing Authority in 1934, which offered government insurance to banks that provided mortgages for newly built homes. As the US economy picked up steam in the decades following the Second World War, its growing middle class moved en masse to newly built suburbs.

The expansion of zoning laws also played an important role in the changing nature of cities. Whereas once they evolved organically, with factories and offices intermingled with residential buildings, shopping and entertainment, a more interventionist approach to urban planning began to take shape in the early decades of the twentieth century. To protect against the ills of pollution and congestion, planners used building permissions to separate cities into distinct zones of activity. Separate places to work, live and play were enshrined in law.

While suburbanization is often associated with the United States, European capitals were not immune from sprawl. In 1900 the population of London's inner boroughs was more than triple the population of its outer boroughs. By the middle of the century, the outer boroughs had already closed the gap, with a deliberate programme of inner city slum clearance and the German bombing blitz accelerating the move to the suburbs. By 1990, the population of the inner boroughs had fallen to around 2.5 million, from a peak of 5 million at the start of the century.[14] In Paris, a commitment to the elegant uniformity of Haussmann's architecture, central to Paris's distinctive image, has limited options for increasing density of the urban centre. All population growth since the early twentieth century has taken place outside the Paris '*périphérique*' ring road.

Suburbanization was a major contributor to economic growth in the post-war decades, especially in the US. Demand for cars, houses, appliances and furniture supported the country's manufacturing sector. By the 1960s, however, factories and office parks had also increasingly relocated to the suburbs to take advantage of the more relaxed constraints on construction and cheaper land. Chicago, for example, lost 211,000 jobs between 1960 and 1970 while its suburbs gained 548,000.[15] The result of the flight from cities was a sharp contrast between the robust growth of the US as a whole and the decline of its inner cities.

ENTRENCHING DISADVANTAGE

Around the time of the First World War, millions of Black Americans in Southern states suffering under the oppressive weight of Jim Crow laws began moving northwards to fill the shortage of manufacturing workers in cities of the Northeast and Midwest – a process known as the Great Migration. In 1900, 90 per cent of the Black population in the US resided in Southern states; by 1970, that figure had fallen to 53 per cent.[16] As the journalist Emmett J. Scott, who witnessed the migration in its early years, wrote: 'they left as though they were fleeing some curse ... They were willing to make almost any sacrifice to obtain a railroad ticket.'[17] Cities like Detroit and New York became the new centres of gravity for the Black population.

In his influential 1987 book *The Truly Disadvantaged*, William Wilson traced how this transformation in the demography of the US set the scene for the collapse of inner city economies in the latter decades of the twentieth century, arguing that 'The social problems of urban life in the United States are, in large measure, the problems of racial inequality.'[18] With suburbanization accelerating in the post-war decades, the Black population increasingly found itself trapped in inner city areas by discriminatory housing practices, as White suburbanites feared the impact of Black neighbours on housing values. The economic prospects for many Black families worsened after these discriminatory housing practices were outlawed in 1968, as the small number of middle-class Black families moved out to the suburbs. The poorest members of the community were left behind just as factories and office parks were disappearing from those inner city areas. The result was a spiralling process of joblessness, increasing concentrations of poverty, persistent social problems, and rising welfare dependency in inner city ghettos. In New York, crime became widespread, the subway became increasingly unsafe and unreliable, and by 1975 the city government was teetering on the edge of bankruptcy.

The declining quality of inner city schools compounded these challenges and contributed to the perpetuation of urban poverty across generations. Well-meaning affluent parents understandably aim to get their children the best education they can, including moving to neighbourhoods with better schools. These parents tend to play a more active role in helping their children with their learning, as well as providing a more stable home environment that assists in child development.[19] By helping their peers directly, and simply by setting an example in prioritizing their learning, the children of these families can help bring up the overall performance of mixed classrooms. Unfortunately, the opposite is also true. When schools become increasingly homogeneous, with poor children learning with other poor children and rich children learning with similarly rich children, the performance gap widens. Policymakers have long been aware of this, but it has proved to be a tough issue to solve. The outlawing of racial segregation in schools in 1954, and later the introduction of compulsory busing within school districts to improve the racial mix of schools, had the unintended consequence of accelerating the exit of White families from inner city areas in favour of predominately White school districts in the suburbs. Desegregation in schooling profoundly altered the lives of Black students who were able to attend mixed schools, with higher rates of high school graduation and a better chance of escaping from poverty.[20] Mixed schools also helped to foster mutual understanding and acceptance among pupils of different races. But the contribution of these policies to further 'white flight' from inner city areas, and the resulting entrenchment of racial poverty, was also real.[21]

This story is not unique to the US. In Britain, settlement of non-White populations from former colonies in the decades after the Second World War led to a similar pattern, as landlords and lenders excluded these migrants from affluent suburban areas.[22] Neighbourhoods like Brixton and Dalston in London became mired in poverty and relative deprivation. The eruption of riots in Brixton in 1981 served as a stark reminder of the trouble plaguing these areas.

THE GREAT INVERSION

Recent decades have witnessed a rapid reversal in the fate of inner cities. Urban centres have seen increasing employment, falling crime and significant improvements in the performance of public services. Warehouses and factories from London's Camden Town to Sydney's Surry Hills to New York's Brooklyn have been converted into upmarket apartments for well-educated (and typically White) professionals. Former working-class housing has been renovated or redeveloped. Trendy cafes and bars have appeared, along with high-end fitness studios and organic food stores. Even Brixton has seen significant gentrification, as this process of urban transformation via an influx of affluent residents has come to be known.

To be clear, the growth in the population size of inner cities has been modest compared to continued suburbanization. In fact, a new ring of sprawl dubbed the 'exurb' has emerged. What has changed is the socioeconomic makeup of the concentric circles of cities. Whereas once the well-off fled for suburbia, today they are heading back to the urban core, while poverty is increasingly moving out to the suburbs. The journalist Alan Ehrenhalt has aptly described this process as 'The Great Inversion'.[23]

The increased attractiveness of urban centres can be seen in the shifting profile of house prices within the concentric circles of cities. One study of the top 20 cities in the United States over time found a major shift in recent decades in the relationship between house prices and distance from the city centre: while in 1980, house prices increased the further away a property was from the central business district, by 2010, that relationship had inverted.[24] The premium on housing in urban centres has further increased since then: median home prices in the five inner boroughs of New York City increased at four times the rate of the rest of the metropolitan area between the start of 2010 and the start of 2020.[25]

What has led highly paid professionals to swap a freestanding house and garden in the suburbs for high-density living in urban centres? Multiple factors have been at work. The reduction in urban pollution in rich countries in the latter decades of the twentieth century thanks to increasing regulation and the decline of dirty industries is one reason. The Thames River that runs through the heart of London was long a source of revulsion for residents of the capital. Back in 1858, industrial, human and animal waste combined with hot weather to produce such a horrible smell it was named 'The Great Stink'. Despite subsequent efforts to improve the condition of the Thames, it remained abhorrently polluted, to the extent that it was declared 'biologically dead' by the Natural History Museum in 1957. Thanks, however, to a multi-decade programme of cleaning and treatment, the river has made an impressive recovery. Cities like Chicago have enjoyed a similar reversal. In recent decades, the removal of lead from fuel and other environmental regulations have contributed to further improvements in air quality in rich countries, though there is still a long way to go.

Another reason for the return of the well-off is the fall in inner city crime. Between 1990 and 2000 the violent crime rate in New York fell by 60 per cent, roughly coinciding with the acceleration in gentrification. Other major US cities including Chicago, San Francisco and Los Angeles enjoyed similar declines in crime rates. Many explanations have been offered for this development, ranging from the conservative (more strident policing practices), to the controversial (legalization of abortion), to the surprising (the removal of lead from motor fuel). We will not assess the merits of these arguments here, other than to point out that the fall in inner city crime has also occurred in cities like London and Sydney, which should suggest some caution in focusing too heavily on US-specific factors. Whether falling crime rates have led to gentrification, or whether gentrification has in fact driven the fall in inner city crime rates, also remains contested. Well-off residents have less to gain and more to lose from participating in crime. One study by a group of MIT economists sought to resolve this ambiguity by looking at what happened to crime rates following a public ballot in Cambridge (Massachusetts) that ended rent control and therefore brought about a rapid surge in gentrification in the area.[26] The fact that the researchers found a significant drop-off in crime after the policy change lends support to the idea that gentrification could be the cause of falling crime, rather than the other way around.

There is also reason to believe that a fundamental reconfiguration in lifestyle preferences has been underway in recent decades, contributing to the growing desirability of living in urban centres. To understand why, we need to first step back in time to the counterculture phenomenon of the 1960s and 1970s that reverberated out from the US to most of the rich world. With its celebration of personal expression and experimental lifestyles, the movement resoundingly rejected the conformity of the post-war era. Urban centres emerged as the focal point for the counterculture, with neighbourhoods like Haight-Ashbury in San Francisco and SoHo in New York serving as focal points. The low cost of living in these urban areas made them well suited to an artistic lifestyle. Perhaps more importantly, they felt alive and authentic, in contrast to the stifling sameness of suburbia.

In fact, conformity was in many cases a deliberate choice in suburban life. Walter Gropius, the German architect who founded the influential Bauhaus movement of the early twentieth century, strongly believed that artistry, functionality and mass production were not mutually exclusive. That principle also applied when it came to people's homes. Gropius believed that 'For the majority of individuals the necessities of life are the same ... Hence it is not justifiable for each house to have a different floor plan, a different shape, different building materials, and a different "style". To do this is to practice waste and to put a false emphasis on individuality.'[27] Anyone who has lived in or even travelled through the cul-de-sacs of suburbia will have seen this philosophy in action.

Urban centres became especially attractive for those whose lives fell outside the conventional social mores. The gay community in particular embraced the inner city as a place not only where they could escape the judgement of middle-class suburbanites, but also where they could find among its density a critical mass of others who shared their sexual orientation. London's Soho, New York's West Village, San Francisco's Castro District and Sydney's Darlinghurst all became magnets for the gay community. Amid rising immigration in the latter half of the twentieth century, inner cities also continued to perform their historical function of bringing ethnic minorities together into diaspora communities, in contrast to the overwhelmingly White suburbs. As a result, inner cities developed a spirit of tolerance and openness that was conspicuously absent from suburbia.

While the counterculture movement collapsed in the mid-1970s, and remained dormant through the heady days of the neoliberal revolution of Reagan and Thatcher, its spirit resurfaced in a curious way in the twilight of the twentieth century. In his insightful and entertaining 2000 book *Bobos in Paradise*, *New York Times* journalist David Brooks coined the term 'bobo' – or bourgeois bohemian – to describe an emerging breed of educated elite that blended the aesthetic of the counterculture with the economic advantages bestowed on them by the transition to a knowledge economy. This is the type of person for whom it would seem vulgar to spend $15,000 on a home cinema, but perfectly natural to spend the same amount on a rustic-looking kitchen.[28] Today you might identify bobos by an integrated suite of Apple products or an Instagram feed full of pictures of yoga retreats in hard-to-reach places.

For the bobo, the mark of success is not a double-garage home in the suburbs, but a loft apartment in an inner city neighbourhood surrounded by creative types who prove their status as an enlightened member of society. Indeed, the lifecycle of gentrification in recent decades has tended to follow a predictable pattern: first comes the artists, then the real estate developers and then the professionals. This same story has played out everywhere from London's Shoreditch to New York's SoHo to Sydney's Newtown. The process of gentrification is not entirely new. Ruth Glass's 1964 work *London: Aspects of*

Change lamented how 'There is very little left of the poorer enclaves of Hampstead and Chelsea ... The invasion has since spread to Islington, Paddington, North Kensington – even to the "shady" parts of Notting Hill.'[29] In recent decades, however, the process has accelerated and extended into many inner city neighbourhoods that once provided affordable homes to those on lower incomes.

The shifting fortunes of inner city areas over the last two centuries was brought to life in an in-depth study of the history of Greene Street in SoHo (New York), led by NYU's William Easterly.[30] Originally serving as agricultural land, the block was developed in the 1820s into townhouses for merchants, doctors and lawyers fleeing the poor sanitation and congestion of downtown in the wake of a series of outbreaks of yellow fever. The introduction of streetcars in the 1830s and 1840s helped fuel the growth of the area by facilitating easy commuting. However, the area fell into disrepute after the 1850s as it developed a growing industry of sex workers serving the hotels and musical halls on Broadway. The development of New York's garment industry led to a new phase in the life of Greene Street from around 1880, with SoHo a favoured area due to its access to both the Hudson River piers and the New York Central Railroad. Many of the small brick houses on Greene Street were knocked down and replaced with factories and warehouses, while the repurposing of the area for industrial usage led to a significant decline in the number of residents. The disappearance of New York City's garment industry in the early decades of the twentieth century led the area to fall into disuse, before it re-emerged as an attractive area for artists and art galleries starting around the 1960s thanks to its cheap rents and large spaces. As SoHo's artistic bona fides made it an attractive target for bobos in the 1990s, the area began to rapidly gentrify, with many of the artists pushed out to areas like Williamsburg (which would later experience its own wave of gentrification). Today, Greene Street is home to luxury apartments and high-end retail, having come full circle over the course of 200 years.

GENERATIONAL CHANGE

The growing demand for inner city living has coincided with the arrival of the millennial generation into adulthood. As a result, the movement of a rising share of those born into this generation back out to the suburbs in recent years has been viewed as a sign by some that the rising appeal of the urban core may be fading. To see why this is not the case, we need to understand the relationship between the city and stages of life.

Curious about the world and hungry for experiences, many young people setting out on their careers are drawn to urban centres. This is the time in a person's life when they may have achieved financial independence – or something close to it – but are not yet constrained by the need for a larger space to accommodate a family. Unbounded by such constraints, these individuals head to the city and all the excitement it has to offer. The role of inner cities as marriage markets, which we briefly noted in the previous chapter, reinforces this. Even in an age of online dating, very few couples choose to conduct their relationships entirely in the virtual realm, at least after the first few encounters. And the further one gets from busy urban centres, the lower the likelihood of finding a good match. As singles couple up and start a family, the pressure for more space increases, while time available for enjoying big-city life shrinks. The result is an exodus from cities for parents with young children.

We can see this lifecycle pattern clearly in the net migration flows in and out of the inner boroughs of London. While many teenagers leave the city after graduating from school to attend university, they return after their degrees are completed, and bring with them many more young adults from across the country. As a result, net migration into inner London flips from negative to positive for adults in their early-twenties, and continues rising into the mid-twenties. After that, it begins to taper off and eventually turns negative again as adults have children and leave for the outer suburbs and commuter towns.[31]

Over the past two decades, however, that story has evolved in two important ways. First, the amount of net migration into inner London by adults in their mid-twenties has nearly tripled. Second, the age at which net migration flips direction – with more adults exiting than entering – has shifted by a full decade, from 34 to 44. The reason for this lies in changing demography. The average age of marriage has increased significantly in the past few decades. When Prince Charles and Lady Diana married in 1981, the average age of first marriage in the UK for women was 22 and for men 24; when Prince Harry and Meghan Markle married in 2018, these figures had risen to 30 and 32 respectively.[32] Over the same time period, the average age of women at childbirth in the UK has risen from 27 to 31.[33] As young people couple off and start families later, if at all, the pull of the urban lifestyle remains for longer. And while increasingly unaffordable housing in cities like London is now pushing out a growing number of young people, many would rather stay.

GENTRIFICATION AND THE COMMUTING POOR

The Great Inversion has taken an immense toll on many of society's most disadvantaged. As wealthy urbanites move in, the existing poor residents are pushed away. For those who happen to own their properties in gentrifying neighbourhoods, this process can create a windfall financial gain. Unfortunately, the most disadvantaged in these areas tend to be renters, who find themselves confronted with rapidly rising housing costs. While the effects of gentrification may be more muted in cases where the neighbourhood in question was once made up primarily of industrial and commercial real estate, the stock of such property has rapidly been exhausted in places like New York, Chicago and London. The result is a combination of increasingly concentrated disadvantage in a small number of inner city neighbourhoods – such as the Bronx in New York or Englewood in Chicago – and a general outward movement of poorer people into the suburbs.

Often, those who are pushed out by gentrification end up far out in the exurbs, where property is cheap but jobs are sparse and commute times to the urban centre are oppressive, especially for those who cannot afford a car and rely on public transport. Sheila James, a public health worker, told the *New York Times* that property prices in San Francisco had forced her so far out of the city that she faced a 2:15 a.m. start on a normal working day.[34] Between 2000 and 2015, the share of neighbourhoods in exurban areas with a poverty rate above 20 per cent more than doubled in the US.[35] And while commute times in the US have been rising across the board, they have been rising far more rapidly for Black and Hispanic workers.[36] Whereas the truly disadvantaged were previously those trapped in poor inner city neighbourhoods, they are increasingly those trapped in low-density areas on the periphery of cities.

The takeover of inner city areas by educated professionals has come at a great price, but it is unclear whether the alternative is any more appealing. As discussed in the previous chapter, the ability of cities to attract high-skill knowledge workers is essential for their success in today's economy. And these workers now want to live in trendy urban centres into their late-thirties and beyond. The fact that gentrification has been slower or missing entirely in struggling cities like Detroit and Cleveland is not a coincidence.

A CITY FOR ALL

What should be done about this? Making today's cities work for all their inhabitants, not just a lucky few, will require three pillars: fairer schooling, fairer housing and fairer public transport. To start with schooling, it is again instructive to look at rich countries that have succeeded in achieving relatively equal outcomes for students irrespective of their socioeconomic background. Japan's schooling system is perhaps best known for the demanding pressures it places on pupils, but it is also one of the most egalitarian systems in the world.[37] The approach to education in Japan offers a number of useful learnings. The first is to decouple the financing of schools from local revenue streams. In the United States, nearly half of school funding comes from local government revenues, which vary heavily depending on the affluence of an area.[38] By contrast, in Japan funding for teacher salaries, school buildings and other expenses primarily comes from national and prefectural governments.[39] The fact that very few primary and lower secondary school students attend private schools in Japan also means that everybody participates in the same education system during these pivotal years.

The second lesson from Japan is to move away from the direct hiring of teachers by schools. Teachers in Japan are hired by prefectures – analogous to states – and typically rotate through multiple schools over the course of their career, especially early on. This allows the government to direct high-performing teachers to disadvantaged areas, rather than teacher quality acting as a mechanism for further widening the socioeconomic gap in school performance.[40]

Many other ideas exist for reducing educational inequality, some easier, others harder. Research by Roland Fryer of Harvard has demonstrated that simply providing increased training for public school principals can have a meaningful impact on student achievement.[41] In Britain, the teachers' union has advocated for a certain number of places in high-performing schools to be ringfenced for disadvantaged students from outside the catchment area.[42] A similar concept exists in the US in the shape of 'magnet schools', which focus on drawing together gifted students irrespective of background, though the impact on the disadvantaged students who are not fortunate enough to make the cut-off is unclear. Perhaps the most powerful mechanism for reducing disparities in educational outcomes within cities is to tackle the challenge of affordable housing, which prevents poorer families from accessing better schools in wealthier neighbourhoods.

In the early decades of the twentieth century, as public concern continued to grow over the quality of life for the working poor in overcrowded and ramshackle inner city housing, many rich countries began major efforts to clear such areas and replace them with social housing, a process that accelerated in the aftermath of the Second World War. Britain was a particularly enthusiastic adopter, building over four million social houses between the end of the war and the start of the 1980s.[43] Such initiatives were not without their critics. For example, in the 1960s the urbanist Jane Jacobs argued forcefully against the prevailing approach in New York City of bulldozing slums and replacing them with lifeless, poorly designed housing projects that undermined local communities.[44] But social housing has played an important role in reducing inequality in cities by ensuring that all residents can have their basic need for shelter met. When interspersed through a city, it has also helped to combat the tendency towards socioeconomic segregation. In London, for example, social housing was also incorporated into wealthy areas like Chelsea or Primrose Hill in the post-war decades, with the notable benefit of disadvantaged families gaining access to the same schools as their wealthy neighbours.

Social housing has since fallen out of favour in many countries. Margaret Thatcher sold off much of Britain's stock to its tenants in the 1980s. France and Germany soon followed suit.[45] Economists began to advocate a more market-friendly model of providing direct financial support in the form of housing vouchers for poor families, further contributing to the increase in socioeconomic segregation in many cities as low-income households consolidated around low-income areas where rent is cheap. In recent years, Britain has sought to remedy this issue by requiring new residential developments above a certain size to also incorporate a number of housing units that are rented out below market rates. Even including these homes, however, the combined supply of social and discounted housing in Britain has been in steady decline since the 1980s.[46]

Cities like Vienna illustrate the alternative. More than 60 per cent of the inhabitants of the city live in subsidized rental housing – compared to just over 20 per cent in London and just over 5 per cent in New York – roughly half of which is owned by the municipal government and the other half by subsidized non-profit cooperatives.[47,48,49] A relatively liberal upper limit on the household income that qualifies for subsidized housing in Vienna – €45,500 ($47,000) for a single person or €68,000 ($70,000) for a couple – means these housing blocks bring together people across a broad swathe of the socioeconomic spectrum.

The rapid increase in house prices in many major cities has made home ownership – and the wealth creation it brings – increasingly unattainable for many. House prices in cities like London, Paris, San Francisco and Sydney have grown much faster than median incomes in recent decades, making it hard to accumulate the savings needed to transition from renting to owning. The correction in house prices underway at the time of writing will not be enough to meaningfully change that picture over the long run. There are two main culprits behind the decline in housing affordability. The first is years of low interest rates, which until very recently made debt unusually cheap. That made it possible for those who already had the initial capital for a deposit to pay more for a home, either for themselves or as a rental investment, pushing up prices and making it harder for those without a deposit to get on the property ladder. The second is the long-term deceleration in the building of new houses. Adjusting for population size, rich countries are now building less than half as many houses as they did in 1970.[50] The problem is particularly acute in inner cities, many of which have seen very little growth in the housing supply in recent decades, with construction activity skewed towards refreshing the existing stock. Between 2010 and 2019, total housing units in New York City's five boroughs increased by just 6 per cent, compared with job growth of 21 per cent.[51]

Cities need to stop growing out and start increasing density. This need not imply endless high-rise buildings and the destruction of heritage architecture – much could be achieved through greater use of mid-rise development and faster conversion of former industrial spaces. As populations in rich countries age and as young people live alone for longer, accelerating the conversion of family homes into single units can also help.

The final pillar for fairer cities is fairer public transport. Access to cheap transportation has long been essential in giving the disadvantaged inhabitants of cities the means to access gainful employment. Yet the systems of public transport that exist in many major cities were often designed and built more than a century ago, at a time when the geography of poverty was very different. As inner city areas become gentrified and poverty shifts outwards, the risk is that transit systems in cities like London or San Francisco end up subsidizing the wrong population. For instance, in London, the cost of a monthly travel pass is highest for those who need to commute from outer zones into the inner city, yet these areas are exactly where poverty is growing the fastest.[52]

Questions also need to be answered around the financing of public transport. In London, tickets fund over 70 per cent of public transport income, twice the share in Paris.[53] Largely as a result, a monthly travel pass in London for zones 1 to 3 – which loosely corresponds to the inner boroughs – at the time of writing costs £174 ($207).[54] That compares to just €75 ($78) in Paris for an all-zones metro system pass that covers a similar distance.[55] Transit systems like London's that rely on ticket revenue more than general taxation place a greater financial burden on the poorest inhabitants of the city and increase the risk of them being trapped inside struggling neighbourhoods. Still, for all its flaws, the London system is far superior to car-centric cities in the US like Los Angeles or Atlanta, where a distinctive lack of public transit options acts as a poverty trap for any resident unable to afford a car.

In the next chapter, we ask if and how this story might change as a result of the surge in remote working following the coronavirus pandemic. Are we at the start of a new cyclical process of well-off workers fleeing urban centres for greener pastures?

5

Remote Work: The Threat to Cities

Enthusiasm for the idea of remote working is nothing new. The first to systematically explore its potential in the modern world was the US physicist Jack Nilles, who in 1976 published a book titled *The Telecommunications–Transportation Tradeoff*.[1] Writing against a backdrop of soaring oil prices and growing excitement about the personal computer, Nilles argued that society was approaching a tipping point. Once the cost of telecommunications equipment fell below the cost of commuting to work, which he forecast it would soon do, centralized offices would become a relic of the past. A few years later, the popular futurist Alvin Toffler echoed this prediction in his 1980 book *The Third Wave*.[2] In a throwback to pre-industrial cottage industries, in which the manufacturing of products like textiles occurred within the worker's own home, he imagined the 'electronic cottage' re-establishing the home as the focal point of work.

The rapid growth of the internet in the 1990s further fuelled the belief that a large-scale transition to remote working was just over the horizon. In 1993, management guru Peter Drucker declared that 'It is now infinitely easier, cheaper and faster to do what the nineteenth century could not do: move information, and with it office work, to where the people are.'[3] When communication is instant and free, and all the world's information is a click away, what need is there for the office? And in a knowledge economy, if there is no need for the office what need is there for the city?

The growth of remote work in subsequent decades was far more modest than anticipated. By 2019, US workers spent just 5 per cent of working time at home.[4] And for most, 'working from home' meant catching up on emails or paperwork in the evening after leaving the office, less a radical reimagining of work and more an intrusion into personal time.

In the space of a few months in early 2020, however, the share of US workers doing their jobs from spare bedrooms and kitchen tables surged to 60 per cent as a result of national lockdowns brought on by the Covid-19 pandemic.[5] Similar rates were seen across other rich countries. While essential workers in frontline service and operational jobs continued to report for duty during national lockdowns, it quickly became apparent that it was technically possible, thanks to advances in recent decades, for white-collar workers to do their jobs without commuting to the office.

Now that offices across the world have largely reopened and the pandemic is behind us, the big question is whether the experience will catalyze a permanent shift towards remote working, or whether we will steadily revert to old patterns. Will cyberspace replace the city as humanity's workshop? Workers have adapted to videoconferencing in place of in-person meetings, and managers have learnt to trust their teams and focus on results rather than time in the office. Were the prophets of remote working simply ahead of their times?

The answer will have profound implications for the forces of urban agglomeration in our knowledge-based economy. As discussed in Chapter 3, knowledge workers are the economic foundation of the major cities of the rich world today. As discussed in Chapter 4, they are also increasingly dominant in urban centres. If remote working becomes the default arrangement for the lion's share of white-collar work, will we see another inversion in the economic geography of cities, and perhaps even their decline?

The last few years have seen much debate over the future of remote work. Most firms have landed now on a 'hybrid' model that involves some time in the office and some time working from home, though the balance varies considerably. Workers are similarly divided. In one global survey in 2021, a quarter of workers never wanted to go into the office again, while another quarter never wanted to work from home again; the remaining half preferred some kind of middle ground.[6]

With time we expect that the downsides of remote working – for employee experience and company performance – will lead many companies to increase the share of days that staff are expected to be present in the office. That will place a ceiling on how far away these workers can move. And many knowledge workers with the financial means to do so have chosen to continue living in urban centres despite a shorter commute no longer offering the same advantage. As argued in Chapter 4, the lifestyle of the inner city, with its cultural variety and ease of access to a wide range of services, is a major factor in the appeal of these neighbourhoods.

Equally, cities have no cause for complacency. The exodus of knowledge workers from inner cities may well have peaked, but the transition to hybrid arrangements will still have significant implications for offices, public transport systems and municipal finances. The answer for cities is not to ignore the trend, but rather to treat it as an opportunity to rethink the way cities are configured to make them fairer and even more vibrant places to live.

HIDDEN COSTS

Commuting to work is among people's least favourite activities. For Britons, it ranks even lower down the list than domestic chores.[7] Commute times around the world have been rising in recent years, as suburbanization continues to push a rising share of urban populations further from the centre of cities. For many, home working has meant more time with family and more rest, not to mention the money saved on fuel or transport fares. It has also meant fewer emissions from travelling: in August 2022, working from home was estimated to be saving 6 billion miles of commuting a week in the US.[8]

But our commutes play an underappreciated role as a psychological barrier between our work and personal lives. Surveys suggest that working from home was an important contributor to the surging rates of anxiety and depression that occurred during the pandemic. In one survey of US workers in 2021, 45 per cent of respondents said they felt working from home negatively impacted on their mental health.[9] In Britain, one poll in 2020 put the figure at 80 per cent.[10] It is difficult to disconnect from a job when you live and work in the same space.

The workplace also forms an important pillar in people's social lives. That is increasingly true as other traditional community institutions decline, a topic we return to in Chapter 6. Globally, more than a third of workers rank a 'good relationship with co-workers' as one of their three most important attributes in a job.[11] Many missed the camaraderie and connection of office life during lockdowns. We may not all work in environments reminiscent of *The Office* or *Parks and Recreation*, but the workplace plays an essential role in combating loneliness and helping to build bridges across our increasingly insular social bubbles. And it is very difficult to create and sustain a sense of corporate community remotely.

There is also good reason to believe that remote work will harm company productivity over the long run. Surveys undertaken since the start of the pandemic have asked workers to rate the impact of home working on their productivity. Such estimates should be taken with a grain of salt. Many respondents attribute productivity gains to the time they get back from commuting. While using the time typically reserved for commuting to plan ahead or catch up on emails might allow workers to get more done in their day, this is extra labour, not greater productivity.

One in-depth study of 10,000 skilled professionals at an Asian technology company before and after the onset of the pandemic offers a richer set of insights on the issue.[12] While remote workers put in more hours, the measures the company used to track worker output – for example, completed segments of code for software engineers – declined. This translated into significantly lower output per hour, which fell by nearly 20 per cent at the upper end of their estimates.

The software used to monitor staff at the company in question filtered out time spent on idle activities such as browsing social media, so we know that declining productivity was not a result of workers simply slacking off while unsupervised. Instead, the authors attributed the declining productivity to two related factors. The first was a significant increase in the time spent in meetings and sending emails. Being physically co-located with colleagues in an office produces a comparatively fluid system of communication. A quick instruction can replace the need for a scheduled meeting. A brief dialogue can eliminate a time-consuming email exchange. The emerging evidence suggests these forms of interaction are difficult to replicate remotely. Uber noted a 40 per cent increase in meetings after the pandemic struck and staff began to work remotely, with the average number of participants attending each meeting rising by 45 per cent.[13] While working from home may give workers more opportunity for distraction-free time on individual tasks, in practice overflowing calendars and inboxes often prevent this.

The second factor behind the decline in productivity identified by the authors was a contraction in the communication networks of workers after switching to remote work. Information is the lubricant of the modern company. To keep the wheels turning, it must flow not only up and down the formal organizational hierarchy, but also through many informal linkages across the business. Sharing a physical office keeps these linkages alive. In his 1977 book *Managing the Flow of Technology*, MIT management professor Thomas Allen demonstrated that the frequency of communication between two individuals declined along an exponential curve the further away in an office they sat, a relationship that would come to be known as the Allen curve.[14] What is particularly striking is how well this finding has held up in the face of all the advances in telecommunications over recent decades. Research suggests that the colleagues we are most likely to email are the ones we regularly interact with face to face.[15] Without the serendipitous encounters the office creates, communication along the informal linkages within an organization dries up.

Highly routine office work may be somewhat shielded from these negative effects. One study of 16,000 call-centre workers in China, for example, noted a 4 per cent increase in calls per minute for workers operating at home.[16] The authors of the study attributed this increase in productivity to a quieter environment at home compared with the office, though workers with children or other dependants may dispute that hypothesis.

At the other end of the spectrum, creative jobs rely heavily on an incidental cross-pollination of ideas that suffers without physical proximity. Steve Jobs was certainly an ardent believer in the value of chance office encounters. When he designed Pixar's offices in 1986, he positioned all the amenities – not just the cafe and mailboxes, but also the bathrooms – in a central atrium to encourage such interactions.[17]

Most office work today falls somewhere in between these two extremes. Over time, however, the continued automation of routine activities will mean an ever-greater share of office work will shift towards the types of activities that thrive on physical proximity. The typical white-collar worker when Jack Nilles was conducting his pioneering research in the 1970s was very different to today's knowledge professional.

The hidden costs of remote working are also likely to fall disproportionately on younger workers who are just setting out on their careers. We might associate apprenticeships with manual trades, but in fact most modern knowledge work relies heavily on sustained informal coaching. It is difficult to replicate the casual feedback after a meeting or the quick corridor question when working remotely, the absence of which particularly harms the learning and growth of younger and newer workers. Moreover, employees who have been in a business for a number of years have already had the opportunity to build relationships and trust with their colleagues through multiple in-person and informal interactions. New joiners, even if they already have some professional experience, lack these networks to draw on. Remote meetings on the whole tend to be more formal, which makes it difficult for junior workers to build the rapport with senior colleagues that is often so necessary to learn and get ahead. Remote work may make work environments more hierarchical, unless care is taken to overcome the divides it can reinforce. This dynamic may also help explain the fact that remote workers are less likely to get promoted.[18] As Nick Bloom, an economics professor at Stanford, has observed, 'You're basically forgotten about. Out of sight, out of mind.'[19]

Part of the problem with remote work is that the technology we have at our disposal remains a poor simulation of physical proximity. The importance of body language and eye contact in effective communication has long been known. A videoconference is at best partially effective in transmitting non-verbal messages, while an email is even more limited. We are only beginning to grasp just how subtle and profound the signals are that we send each other during in-person encounters. In one extraordinary study, a group of researchers at University College London found that the heartbeats of theatre audiences synchronize during live performances, in a similar fashion to what is observed when romantic couples sleep beside one another.[20] Humans evolved to develop relationships through in-person interaction. Replicating this virtually is not yet possible. Many companies are certainly trying. Technology start-up PORTL has developed a hologram screen that accurately reproduces a life-size 3D image of a person. But, at $60,000 a piece, it remains prohibitively expensive for general use. It is possible that one day the challenges inherent to remote work may be engineered away, but for now we believe remote methods of communication will remain primarily a complement to in-person interactions, not a substitute.

Of course, none of this implies that workers need to be in the office five days a week. Some workers and some companies have opted for the binary extremes of fully remote or fully in-person, but we expect hybrid arrangements will continue to solidify as the preferred model, as companies look to find the right balance between the upsides and downsides of working away from the office. What will this mean for cities?

HYBRID CITIES

During the early months of the Covid-19 pandemic a significant share of office workers fled to the suburbs. This led a number of commentators to predict that the shift to hybrid working would lead to a permanent flight from inner urban areas in favour of the suburban and exurban periphery. If a worker only needs to commute to the office a few days a week, the optimal distance that balances the cost and inconvenience of commuting with the money saved from lower house prices further from the city shifts the balance in favour of the suburbs, or so the theory goes. However, while suburban house prices have increased by more than urban ones since the onset of the pandemic, the impact seems to have tapered off. Analysis by UBS, an investment bank, shows that permanent migration out of urban centres in favour of suburban and rural areas in the United States, which spiked during the pandemic, had petered out by the end of 2021.[21] High-frequency house price data from Redfin supports this finding, with the wedge in house price growth rates between inner cities and lower-density areas that emerged at the start of the pandemic disappearing by mid-2022.[22]

Data on how public transit use in cities has changed since the pandemic offer clues as to what has happened. The total number of trips on the London Underground in February 2022 was roughly 60 per cent of its pre-pandemic level.[23] But weekend trips were back to above 80 per cent of their former levels, with travel between 10 p.m. and 11 p.m. during the week nearly fully recovered.[24] In the US, office occupancy rates were still below half their pre-pandemic levels in June 2022, but dinner bookings through the OpenTable platform were back up at 90 per cent of pre-pandemic levels.[25] Once-bustling office districts in places like San Francisco remain far quieter than before the pandemic, but the social life of these cities has quickly regained its vitality.

The lifestyle that the inner city offers is keeping most white-collar workers from moving further away, and its attractions have retained their appeal. Those that did make the switch to the periphery of cities – and drove the temporary wedge in house prices – were probably those who were already at the age boundary at which suburbanization starts to occur. The pandemic may have brought forward their move, but the longer-term pattern of rising demand for an inner city lifestyle is not going away.

When it comes to offices, however, there is reason to anticipate substantial and lasting disruption from hybrid working. While the share of the US labour force working remotely has fallen from its lockdown highs, it remains well above pre-pandemic levels, and seems likely to stay there. As of August 2022, nearly 30 per cent of workers were still remote at least one day a week.[26] Some companies are seizing the opportunity to cut back their property portfolios. Others are keeping their offices but investing in transforming them. Partly that is about redesigning the way space is used to focus more on collaboration, and partly it is about jazzing up offices in a bid to draw otherwise reluctant workers back in. A number of companies are also taking the opportunity to transition to offices with a lower environmental footprint. Google parent Alphabet is spending $1 billion to purchase its current central London office, which it plans to refurbish and continue using alongside a new London headquarters that is nearing completion.[27] Both buildings will prioritize flexible workspaces to help foster innovation and creativity. The move follows a similar decision during the height of the pandemic in 2021 to purchase a Manhattan office building for over $2 billion.[28] Citigroup has also committed to a major refurbishment of its EMEA headquarters in London's Canary Wharf, with a greater emphasis on collaborative workspaces and the seamless integration of technology,[29] while Salesforce is redesigning its headquarters in San Francisco around collaborative 'community hubs' in place of rows of desks.[30]

The gap between the top-quality offices, boasting sleek designs, first-rate facilities and green credentials, and the rest of the ageing office stock is widening rapidly, a fact reflected in increasingly bifurcated office rental and sales prices.[31] Businesses are upping their office-space game, partly to encourage people back to work and partly to stand out in the war for talent. Foosball and table tennis don't cut it any more. Many offices today have in-house baristas, great food, and craft beer in the evenings. Google's new London offices will include a swimming pool and rooftop running track. With time, offices may come to be seen as places to connect and socialize more than places to find a desk. Zoom, admittedly with a vested interest, argues that offices are 'no longer for doing good work so much as making good work possible, facilitating collaboration and fostering camaraderie. The office is now like an off-site.'[32]

Early in the pandemic, some suggested that a 'hub and spoke' model, with satellite offices located closer to workers' homes, would emerge as the new standard configuration.[33] Increasingly, it looks like workers' homes will be the 'spokes', while the 'hubs' will be cutting-edge office buildings in prime urban locations.[34] It is in such spaces, which bring together workers from a variety of teams, that the cross-pollination necessary for the vitality of modern companies will take place. If anything, satellite offices in less central locations will decline in importance, as they offer neither the convenience of home working nor the diversity of interactions possible in central offices.

The overall result is likely to be both a decline in the amount of office space required and a 'flight to quality', with demand shifting from older, lower-grade buildings to newer, premium ones. Office landlords will be forced to either convert or upgrade their properties to avoid being stranded with 'zombie' buildings, particularly as more stringent environmental regulations begin to come into force in the years ahead.

As a result of the shift to hybrid arrangements, public transit usage is also likely to remain below its pre-pandemic peaks for the foreseeable future, especially during rush hours, while retail spending around offices will suffer a sustained reduction. The resulting impact on the small business ecosystems of central business districts, from lunch spots to dry cleaners, should not be underestimated. Nor should the negative impact on the sustainability of municipal services. In the US, cities raise around 60 per cent of their own tax revenue.[35] The key revenue sources cities rely on – property taxes, sales taxes and income taxes – will all be heavily impacted by the shift to hybrid working. The financial viability of public transport systems, whether funded by tickets or local taxation, will also need to be addressed. Yet there are, equally, opportunities to be found.

CHALLENGE OR OPPORTUNITY?

Canary Wharf in London is not generally seen as an exciting place to spend time. Towering office buildings, home to banks and other financial institutions, dominate the landscape. Prior to the pandemic, a mass of suits would flow in and out of its underground tube station every day, leaving the area largely dead after about 7 p.m., and deserted on weekends. It is a classic example of what urban planners describe as a single-use area. Canary Wharf is a stark contrast to the more vibrant Shoreditch roughly three miles away, that melds together residential living, a heaving nightlife and a thriving tech cluster in what urban planners refer to as a mixed-use area.

As discussed in Chapter 4, the prevailing urban planning philosophy that took hold in the twentieth century entailed separating different areas of the city into different functional uses. This kind of separation will become increasingly untenable in a world of hybrid working. Business districts in particular must reimagine themselves for a working life less centred around the office. Canary Wharf, for example, is not only converting office buildings into flexible working spaces and constructing one of the largest commercial research labs in the world, but also building thousands of apartment buildings in an effort to redefine itself as a mixed-use area.[36,37] Such efforts should be actively encouraged, given their potential to help ease pressure on house prices in urban cores where demand for housing has far outstripped the slow pace of construction in recent decades.

The transformation of office buildings into apartments is already showing signs of momentum. In 2021, 41 per cent of apartment conversions in the US were from former offices.[38] Such adaptations are not always straightforward – many office buildings lack the level of natural light needed for residential usage – but there are promising signs that developers are finding clever workarounds. In New York's Financial District, the conversion process is already well advanced.[39]

Converting soulless business districts into vibrant mixed-use areas requires a long-term commitment that should not be underestimated. Phoenix has been trying for decades to convert its central business district into a thriving downtown that combines office spaces with residential living and cultural life, but it has been slow going. High-end condominium towers that were built in the early 2000s in an effort to bring residents (and their wallets) to the area have struggled to gain traction.[40] Persistence – helped along by the construction of a light-rail system and the development of Arizona State University's downtown campus – is now starting to pay off, with rising numbers of residents supported by trendy cafes and restaurants and a thriving retail scene.[41] Transformation is possible, but it requires coordinated effort and investment across many years in order to reset the image of urban areas that inhabitants associate with work and not play.

The shift towards hybrid working also presents an opportunity to bring the urban back into suburban. As knowledge professionals spend less time in central business districts and more time in their neighbourhoods, they will bring with them more demand for trendy lunch spots, fancy gyms and other such services. Right now, the typical place to find these things is in gentrified inner city neighbourhoods, but already some suburbs are beginning to reinvent themselves: the term 'hipsturbia' has recently emerged to describe young and vibrant mixed-use districts in suburban areas.[42] If they can pull off the transformation, suburbs will not only benefit from increased demand, but will also help to ease the pressure on inner city neighbourhoods that have become unaffordable to all but the highest paid.

Long a symbol of the suburban lifestyle, the reinvention of malls offers an illustration of what is possible. The first indoor mall in the

United States, the Southdale Center, was opened in Minnesota in 1956. More than simply a shopping destination, it was imagined by its designer, Victor Gruen, as a community space that would bring together the inhabitants of suburbia, and for a time it succeeded in doing exactly that.[43] But the uniformity and overtly materialistic experience of malls is not in line with the bobo sensibility we described in Chapter 4. Moreover, the rise of online commerce has undermined the viability of many of the department stores that served as anchor tenants for malls. As a result, this mainstay of suburban life has been in rapid decline, with many malls largely or completely abandoned. Pioneering efforts are demonstrating that such spaces can be reimagined for changing lifestyle preferences. The Belmar redevelopment in Lakewood, a suburb of Denver, illustrates this. Built on the site of the sprawling Villa Italia mall that closed its doors in 2001, the mixed-use development is a thriving collection of cafes and restaurants, residences and office spaces, and has helped put Lakewood back on the map. House prices in the county are up more than 80 per cent over the past decade.[44] For workers spending a significant share of their time working from home, such redevelopments offer a significant increase in the available supply of attractive neighbourhoods to choose from.

Whether or not suburbs can form viable clusters of employers to ease pressure on central business districts is another matter. While developments like Belmar have attracted corporate tenants, much of their office space is taken up by the non-tradeable services we described in Chapter 3, such as physicians' offices, legal practices and realtors. The general trend in recent decades has been away from suburban office parks, with companies like General Electric and McDonald's moving their headquarters from the suburbs back into, respectively, the inner cities of Boston and Chicago.[45] The shift to hybrid working, which allows a greater share of routine interactions with customers and suppliers to happen remotely, may help widen the circumference of locations in which companies are able to operate without losing the benefit of proximity to business clusters. And if suburbs can replicate more of an urban lifestyle for their residents, they will be better positioned to draw in the knowledge professionals that companies are eager to locate near to. Suburbs such as Arlington in the Washington DC metro area show that this is not a pipedream. By establishing itself as a knowledge hub for cloud computing, artificial intelligence and cybersecurity, centred around mixed-use neighbourhoods like National Landing, Arlington is emerging as a key pillar of the wider metro economy.[46]

Making this work will require new ways of thinking about transportation. Rather than a radial system that is oriented around taking people in and out of a single central economic zone, dispersing economic activity across the wider metro area will require a more complicated web of connections. This adds further urgency to the need for reimagining public transit layouts for the changing economic geography of cities, consistent with the dispersal of access we recommended in Chapter 4.

The growth of remote working also offers new opportunities for cities that currently lack the supply of knowledge professionals needed for success. While workers on hybrid schedules will not be able to decouple themselves entirely from established knowledge hubs, those who work for companies that have made the decision to shift to fully remote working will have much greater latitude. A good share of these workers will choose to stay in cities like London or San Francisco for the lifestyle they offer, but others will look elsewhere. The question is: where will they go? Places with beaches, such as Hawaii or Cornwall, are proving popular, as are places with access to the great outdoors. Cities like Austin, which offer both an attractive lifestyle and an already-established community of high-skill workers, have done particularly well, perhaps in part because remote workers do not want to move somewhere where they will miss out entirely on job opportunities that require them to come into the office a few days a week. It's notable that house prices in Austin rose 65 per cent between January 2020 and January 2022, but in Santa Fe – a city with a thriving cultural life but not many high-skill jobs – they are up only 28 per cent, less than the national average of 31 per cent.[47]

The rise of remote work, even though only partial, presents an opportunity to reset the trajectory towards excessive centralization of economic activity and concentration of wealth in a small number of central neighbourhoods in a small number of cities. The rise of remote working fits a broader pattern of an increasing share of our interactions shifting from the physical world into cyberspace. In the next chapter, we explore what this is doing to the bonds that tie our societies together, and the role that cities must play in healing the growing rifts between us.

6

Cities, Cyberspace and the Future of Community

Throughout history cities have played a central part in bringing people together in the continual interactions that breed a sense of shared identity. Cities like Uruk or Athens provided the public spaces in which citizens could trade, converse and forge a community.

What if technology made it possible to recreate this power of proximity at a global scale? What if all of humanity could be brought into a single network of communication and information? Could such a technology make cities' ancient role as loci of community redundant, and allow people to finally come together and stop defining the world in terms of 'us' and 'them'?

Not too long ago, many had such utopian hopes of the internet. In 1993, the writer Howard Rheingold published *The Virtual Community*, in which he described with breathless enthusiasm the emergence of The WELL – short for Whole Earth 'Lectronic Link – a type of embryonic social network that allowed members to connect with one another over the internet and build shared interest groups. In 1996, journalist and cyber optimist John Perry Barlow published his *Declaration of the Independence of Cyberspace*, in which he described his vision for the internet as 'a world that all may enter without privilege or prejudice accorded to race, economic power, military force, or station of birth … where anyone, anywhere may express his or her beliefs, no matter how singular, without fear of being coerced into silence or conformity'.[1] The fact that key figures like Stewart Brand, who founded The WELL, were also central participants in the 1960s counterculture, with its egalitarian and universalist ethos, helps explain some of this utopian hyperbole.

Far from societies coming closer together, the last few decades have seen many of them unravelling. In the United States, where the General Social Survey provides consistent longitudinal data, the share of adults who agree that most people can be trusted has fallen from nearly 50 per cent in the early 1970s to around 30 per cent today.[2] And while back in 1960 just 5 per cent of Republicans and 4 per cent of Democrats would have been unhappy to see their child marry a person from the opposite political party, the equivalent figures in 2010 were 49 per cent and 33 per cent.[3] Conspiracy theories on the right and 'cancel culture' on the left make constructive dialogue nearly impossible. And while the US may be the most extreme example of the division, Britain, France and many others are also finding themselves increasingly fractured.

The internet and social media cannot bear all the blame for a trend that has been underway for some time. Back in 2000, long before Facebook, Twitter or Reddit came into existence, the sociologist Robert Putnam released his ground-breaking book *Bowling Alone*, in which he traced the declining participation in the voluntary associations that had historically played such an important role in building trust among citizens from all walks of life.[4] Too much time in front of the television, and to a lesser extent growing commute times as a result of urban sprawl, were the main culprits in that decline, not the internet.[5] And political partisanship in the US has been deepening since at least the mid-1990s when Newt Gingrich took over as House Speaker and shifted the tone of political debate.[6]

What the internet has done is pour fuel on the flame. By allowing people to filter what they see and with whom they interact, social media has increasingly broken down the bridges of shared experience and mutual comprehension on which the healthy functioning of society relies. At a time of rising inequality and disillusionment with the status quo, the consequences are perilous. Far from bringing people together and finding common ground, as was hoped in the 1990s, the internet has exacerbated divisions.

In this chapter, we will argue that the development of the internet has made cities more important than ever as a means to overcome divisions. But cities must evolve if they are to rise to this potential.

NEW TECHNOLOGY, OLD IDEA

The vision of the cyber optimists – that advances in the technology of communication would allow humanity to bridge its differences and come together as a unified whole – was not entirely original. In the 1960s, philosopher Marshall McLuhan was predicting that the connection of the world through increasingly sophisticated electronic modes of communication would eventually lead to the creation of a 'global village' in which everyone would share in a collective identity.[7] In 1881, *Scientific American* predicted that technologies such as the telegraph foreshadowed 'a day when science shall have so blended, interwoven, and unified human thoughts and interests that the feeling of universal kinship shall be, not a spasmodic outburst of occasional emotion, but constant and controlling'.[8]

But shifts in the technology of communication often change societies in unexpected ways. When the printing press was first introduced in Europe, the Catholic Church expected it to unify Christendom by quickly and accurately reproducing a single authoritative version of scripture.[9] In practice, the printing press made it possible for iconoclastic thinkers such as Savonarola to challenge church authority, fuelling the hysteria that led to the Bonfire of the Vanities, in which books and artworks were burned in a fit of zealotry. Ultimately, by fuelling the rise of protestantism, the printing press cleaved Christianity in two. The new technology allowed the Bible to be read directly in people's own languages, and made possible the rapid dissemination of Luther's sermons. The volume of Luther's texts published during his lifetime exceeded those published by all his Catholic opponents put together five times over.[10] Later, as religious belief began to wane following the Enlightenment, the printing press helped to reorient the populations of Europe around nationalist identities as publishers standardized texts using primary dialects, and newspapers contributed to instilling a sense of a unified people defined by territorial borders.[11]

Radio and television helped to further cement these unified national identities by broadcasting content across a much wider space. With the advent of the radio, millions of people were tuning in to the same source of entertainment, listening to programmes like *Amos 'n' Andy* in the US or *The Archers* in Britain, as well as national news broadcasts. Television extended the same pattern, with shows such as *The Adventures of Ozzie and Harriet* helping to create a shared idea of 'normal' life in America, and the nightly news provided by a few major channels emerging as a national ritual. In 1970, three broadcasting networks and their affiliates represented 80 per cent of all viewing in the United States.[12]

These examples suggest that technological advances that increase our ability to disseminate ideas can have very different impacts in different contexts. On the one hand, they make it possible for a wider group of people to embrace a set of shared stories and create what Benedict Anderson has referred to as imagined communities. On the other hand, they make it possible for alternative narratives to gain a foothold, for better or for worse.

The internet is, of course, not merely a technology for transmitting our ideas to the world. It is also a tool for seeking out others and engaging in dialogue. And it is not the first such technology to do this either. In the late nineteenth and early twentieth centuries, the telephone enabled instantaneous and lifelike communication between households for the first time in history. Predictions of a profound rewiring of human relations followed.[13] In reality its impact was modest: rather than building new connections, the telephone merely allowed people to converse with the same people more frequently, and to better maintain relationships with existing friends and family across distances.[14]

The internet, and in particular social media, has characteristics of both the television and the telephone. It makes it possible for anyone with a laptop or smartphone to cheaply and instantly transmit their ideas, and for anyone to respond or relay them. But at the same time, it gives people the power to tune in to only the networks of their choosing. That combination makes for a dangerous catalyst in a divided world.

CYBERBALKANIZATION

Contrary to popular opinion, there is little evidence that social media undermines 'real life' relationships. Social media use is actually positively correlated with the size of close personal networks.[15] This is probably because more sociable people tend to be more active social media users, using the technology as an efficient way to keep up to date with the lives of their friends and more readily organize in-person contact. There is a growing body of evidence that social media use among teens contributes to anxiety and depression, as they find themselves in a state of constant comparison with their peers.[16] But that is an argument for age-based limitations, not an indictment of the technology as a whole.

The corrosive effect of social media on the functioning of democracy in rich countries, nevertheless, is becoming increasingly clear. To understand why this is happening, it is helpful to distinguish between what Robert Putnam termed 'bonding' and 'bridging' social capital, where the first refers to the strength of ties *within* a social group and the second refers to connections *between* different social groups. By allowing people to communicate more regularly within their network, and to follow the big and small events that happen in the lives of those they are connected with, social media facilitates bonding. Moreover, social media makes it possible for people to create and sustain communities of like-minded individuals with whom they have greater potential for bonding in the first place.

The trouble comes from the impact of social media on the bridging ties that are essential for reaching outside established networks and connecting unalike people together. Even in the optimistic days of the 1990s, some commentators foresaw this possibility. In 1997, Eric Brynjolfsson and Marshall Van Alstyne, of MIT and Boston University respectively, coined the term 'cyberbalkanization' to describe an alternative vision for the internet in which users retreat into narrow and isolated subcommunities, screening out contact with people who are not like them.[17] Cass Sunstein of Harvard built upon this idea when he described how the self-selection of contacts and sources on the internet could lead to the creation of 'information cocoons' that would continuously reinforce users' pre-existing beliefs.[18] Both predictions have proved to be prescient. As British poet and musician Kae Tempest has aptly put it, 'The internet makes it possible for like people to find each other, and this is extremely important, but it makes it difficult for unalike people to contact each other without their defences up.'[19]

It is important to acknowledge the nuances in the evidence around the impact of social media on society. Abuse and harassment are undoubtedly widespread on the internet, especially when it comes to political discussion.[20] But this does not necessarily mean the internet makes us less civil. Instead, the evidence points to a selection effect in which those who are highly active in political debates on social media are more likely to hold extreme political views and more likely to be of a combative disposition more generally.[21,22] In the words of writer Adam Grant, 'The internet doesn't turn people into trolls. It just makes their trolling more visible.'[23]

In a similar vein, a study that tracked Dutch social media users over time between 2014 and 2020 found evidence that political polarization was more likely a cause of rising social media use than the other way around.[24] Yet algorithms that recommend content based on our prior online activity do act to reinforce our existing beliefs, exacerbating the problem. In another experiment, researchers found that increased exposure to political content on

YouTube further deepened partisan attitudes and increased negative emotions towards candidates of the alternative political party.[25] As mentioned earlier, social polarization was underway well before modern social media. Yet algorithms that recommend content based on our prior online activity do act to reinforce our existing beliefs, significantly exacerbating the problem.

The commercialization of our interactions online is part of the problem here. While the public square of the city had no agenda of its own, the same cannot be said of social media platforms. Facebook and Twitter posts that contain negative comments about alternative political parties generate significantly more engagement.[26] The more user attention these platforms capture, the more advertising revenue they are able to generate, which sets the scene for a clash in incentives between these companies and the societies in which they operate.

It is high time for reforms to social media. Sensible rules can limit the toxicity of online discussion without undermining the ability of these services to create meaningful connections between people and remain economically viable. Capping the number of users who can re-share content before it needs to be copied and pasted in a new post would help to curb the spread of toxic content, a sensible suggestion made by Facebook whistle-blower Frances Haugen.[27] Also useful would be a requirement for social media platforms to verify users, which would not only help to eliminate bots but also disincentivize the most extreme forms of online abuse, including death threats and hate speech.[28] Making social media and other online content platforms legally liable for the content they propagate in the way that other publishers are would be another sure way to quickly curb fake news and hate speech.

Reform of the internet and social media is necessary and overdue, but far from sufficient. It would help slow the deterioration of civic life, but would be unlikely to restore it. It leaves unresolved the question of how to foster bridge building across divided communities. To some, the answer lies in moving an even greater share of our lives into the virtual world.

INTO THE METAVERSE

The notion of a metaverse – a constellation of interconnected virtual worlds in which people can not only work but also play games, shop, attend concerts and more – has surged into public consciousness thanks to major investments by a number of large internet companies. Most notable has been Meta, the company formerly known as Facebook, which is seeking to redefine itself not as a social media company but as a metaverse builder.

Optimists believe that, in contrast to the echo chambers of current social media platforms, the metaverse offers greater scope for bringing society together, thanks to its ability to forge connections across multiple contexts. People with starkly opposing political views, or from very different socioeconomic backgrounds, can find common ground in their shared enjoyment of the same video games or music, for instance. In this way, the metaverse may help to develop what sociologist Ray Oldenburg described as 'third places' – contexts outside of home and work – that bring people from different worlds together in a common community life.[29] And, unlike the real world, space in the metaverse is infinite.

Imagining life in the metaverse takes us squarely into the realm of science fiction. In fact, it is in science fiction that we first see mention of the term. Neal Stephenson's 1992 book *Snow Crash* describes a dystopian world in which the free market reigns supreme in all areas of life, most of society lives in destitution, and those workers who can afford the technology escape the dreariness of their lives by disappearing into the alternative reality of the metaverse.

We have shown in this chapter that predicting the impact of new technology on society is a fool's errand. Only time will tell whether the metaverse lives up to the optimists' vision – or even gains traction at all. The danger is that society uses the new virtual worlds it has created to escape from its problems, rather than confront them. We are physical beings living in a physical world, and we cannot simply let that world decay while we wait for a digital utopia that may never arrive. It is in cities, not in virtual reality, that we will find the greatest hope of mending our bonds, if we can rise to the challenge.

CITIES AS BRIDGES

The modern city is very different to Uruk or Athens. Those cities are merely towns by today's standards. In a city with one million, ten million or even more inhabitants, a resident can only know a tiny fraction of the people he or she lives around and travels past each day. Mark Twain once described New York as 'a splendid desert – a doomed and steepled solitude, where a stranger is lonely in the midst of a million of his race'.[30] Cities are deeply isolating for too many people. Nevertheless, the potential for the modern city to be a force for cohesion is immense.

For a long time, prevailing wisdom among sociologists was that the modern city was the antithesis of a close community. Sociologist Louis Wirth was the first to formalize this idea in an influential 1938 essay titled 'Urbanism as a way of life'.[31] Wirth argued that the transition of human society to the city risked rewiring human relations to be 'largely anonymous, superficial, and transitory', with the individual losing 'the sense of participation that comes with living in an integrated society'.[32]

It wasn't until the 1980s that this belief in the harmful effect of cities on community ties began to face serious challenge. The sociologist Claude Fischer's 1982 book *To Dwell Among Friends* was the opening shot in a new, more positive view of the potential for large cities to help build the connective tissue of modern society.[33] Fischer developed what he called the subcultural theory of urbanism, which centred on the observation that the scale of cities allows individuals to better seek out and forge deep connection with people like themselves. The real magic of the city then comes from its ability to knit these subgroups together into a connected whole, through the power of the serendipitous encounters and shared experiences that only a city can create.

The constant exposure to diverse subcultures in high-density areas also makes residents more tolerant of the values and practices of people different to themselves. What may feel unfamiliar and threatening when encountered the first time gradually begins to feel normal with each new exposure, and with time can even come to be seen as a valuable contribution to the rich tapestry of a varied life.

Not all urban environments are equally effective at achieving this ideal. Suburban sprawl in particular poses a challenge to the connecting power of the city. Lewis Mumford once described suburbanization as 'a collective attempt to live a private life'.[34] In trading proximity for space, the suburbanite also trades serendipitous connection for privacy. And time spent commuting limits our ability to engage with our community – ten minutes extra daily commuting time reduces involvement in community affairs by an average of 10 per cent.[35] That is why expanding mixed-use development throughout the city, as argued in Chapter 5, will be essential in restoring community bonds. A world in which people shuttle between their homes and their places of work and back again offers little scope for forging new connections.

Creating spaces where members of the community can meet is also essential. This has many dimensions, from neighbourhood parks and children's playgrounds to community centres and pedestrianized streets. Simply widening pavements to allow cafes and restaurants to spill onto streets can help. Many neighbourhoods transformed during Covid-19, when indoor seating was prohibited. That is a good start, but much more needs to be done.

Equally, well-connected communities of identical people will not help the challenges described in this chapter. Cities need to ensure that the rich and the poor do not occupy mutually incomprehensible physical worlds. That requires reforming education, housing and transportation to make neighbourhoods more accessible, and investing in struggling cities that trap many of their residents in poor communities. Government support should also be offered to rebuild and modernize the voluntary associations that historically played such an important role in connecting people of many different backgrounds within and across cities.

Last, cities need to find ways to harness technology for connecting people whose lives might otherwise not overlap. Not all social media platforms are the same. Services like Meetup, which connects people based on shared hobbies, or Nextdoor, which acts as a social network and information hub for neighbourhoods, illustrate the power of technology to tie together disparate social groups. One survey found that such applications can increase the portion of residents who feel connected to their local community by 70 per cent.[36] Cities should partner with such services to help improve adoption rates.

There are reasons to be hopeful. In one district of Paris, a group that calls itself the République des Hyper Voisins (or 'Republic of Super Neighbours') is seeking to restore a lost sense of community to the city. One day in 2017, it laid out a 215-metre-long table with 648 chairs along a local street and invited neighbourhood residents to simply come and share a meal together – that most ancient of community rituals.[37] Since then, it has continued to organize community events, bringing together thousands of residents from many different backgrounds. It may not address the structural challenges of sprawl or socioeconomic segregation, but it does prove that we have not lost our yearning to connect with those around us in ways that cannot be achieved in cyberspace.

The social rifts that plague many rich countries today have taken hold over a number of decades. There are no easy solutions, and the problem will not be solved overnight. But with shared commitment and the right investments, cities can play an important part in the healing process we desperately need.

7

Beyond the Rich World

In 1950, Dhaka in Bangladesh was a town of barely 300,000 residents. Old photos from the era show quiet pedestrian streets lined by trees, wide open spaces and a largely undeveloped riverfront.[1] Today, the city is a megalopolis of over 20 million people, with close to 1 million additional inhabitants added to the population every year.[2] It is both densely populated and physically sprawling, steadily drawing the surrounding countryside into its urban fabric.

The speed and scale of urban growth in developing countries is historically unprecedented, far outpacing what occurred in rich countries. London grew by 100,000 inhabitants a year through the 1890s, and New York by around 200,000 in the 1920s.[3] Today, over 30 cities across the developing world are adding more than that every year.[4] From Lagos to Lahore and Shenzhen to São Paulo, the cities of the developing world are booming. A century ago, only one in four of the world's city dwellers lived in poor countries.[5] Today it is three in four.[6] Bar a few exceptions, the cities of the developing world now dwarf those of rich countries.

Part of the reason is population growth. While the population of rich countries has roughly doubled over the past 100 years, it has increased nearly six-fold in poor countries.[7] Dramatic improvements in life expectancy have been pivotal, as the adoption of modern medicine and improved public health practices have curbed the impact of diseases like measles, malaria and cholera. Rapidly falling infant mortality rates are the biggest contributor to this, though longevity has also been improving. In 1920, the average lifespan in India was a startlingly low 25 years, less than half that of Britain at the time. Today it is close to 70 years.[8] Even countries that remain very poor, like the Democratic Republic of Congo, have experienced sharp rises in life expectancy. And while fertility rates have declined in poor countries, they have not done so with sufficient speed to offset rapid gains in life expectancy, particularly in the small number of countries including Nigeria, Ethiopia, the Democratic Republic of Congo and Pakistan that are driving global population growth.

But population growth is not the only factor behind the dramatic rise of cities in poor countries. Over the past century, the *share* of people in developing nations that live in cities has increased nearly ten-fold, and is now above 50 per cent,[9] sharply outpacing the transition that occurred for rich countries during industrialization. It took the United States roughly 60 years to progress from a 20 per cent to a 50 per cent level of urbanization. Central and South American countries took about 50 years to double their rates of urbanization. It took China some 30 years to achieve the same.[10]

Two hundred years ago, a traveller seeking the most awe-inspiring city in the world would have started in China, and in particular its capital Beijing. Victor Hugo, author of *Les Misérables*, never made it outside Europe, but he wrote of the reputation of Beijing's Summer Palace.

> Build a dream with marble, jade, bronze and porcelain, frame it with cedar wood, cover it with precious stones, drape it with silk, make it here a sanctuary, there a harem, elsewhere a citadel, put gods there, and monsters, varnish it, enamel it, gild it, paint it, have architects who are poets build the thousand and one dreams of the thousand and one nights, add gardens, basins, gushing water and foam, swans, ibis, peacocks, suppose in a word a sort of dazzling cavern of human fantasy with the face of a temple and palace, such was this building.[11]

In 1860, during the Second Opium War, the Summer Palace was tragically sacked by the British and French. By this point, a shift in the global urban centre of gravity was well underway, with the cities of Europe and North America growing rapidly on the back of industrialization. By 1900, nine of the ten largest cities in the world were in Europe or the United States, up from just four a century earlier.[12] Today, not one makes that list.[13] China, which once again is home to the world's largest urban population, has at least 160 cities with more than one million residents and more than three times the number of city dwellers as the United States.[14]

Cities lie at the heart of economic development, and no country has ever become rich without urbanizing. Only cities can create the types of jobs that are capable of lifting a large share of the population out of subsistence. And while rich countries can afford to extend services like education, healthcare and infrastructure into remote areas, poor countries cannot. For an economy to transition from poverty to prosperity a large share of the population must move to cities. Plenty of research demonstrates a positive correlation between a country's level of urbanization and its average income.[15]

As we will show in this chapter, however, the story is nuanced. Economic development requires urbanization, but urbanization can and does occur without economic development, resulting in very large amounts of urban poverty. We will explore what separates the success stories, and what must be done to support poor countries that are struggling with persistent urban poverty.

LADDER TO PROSPERITY

Arguably the greatest success story in economic development since the industrial revolution has been the transformation of much of East Asia out of poverty. The region's three major economies – Japan, South Korea and China – have all followed a rapid development path, and urbanization has played a starring role in the process.

Consider first Japan. While economic modernization began in the latter decades of the nineteenth century, Japan in the 1930s was still a predominately rural economy, with a GDP per capita of only about 40 per cent of that prevailing in the US and Britain at the time.[16] After the devastation of the Second World War, however, the country undertook a rapid process of industrial development, which by 1990 had seen income levels rise to 80 per cent of those in the US and 115 per cent of those in Britain, with an urbanization rate typical of rich countries.[17] By the late 1980s, economists were talking reverentially of 'The Japanese Miracle'.

The heart of Japan's rapid development was an economic model centred on manufacturing technologically advanced products – machinery, automobiles, consumer electronics and the like – primarily to export to rich countries. Initially, the Japanese government focused on the development of three major urban areas – Tokyo, Osaka and Nagoya – as the engine rooms of this production, referred to collectively as the Pacific Belt Zone.[18] All three cities had been heavily damaged by bombing during the Second World War, leading to both physical destruction and horrific loss of life. With the support of reconstruction aid, however, Japan was able to rebuild these cities on a foundation of modern infrastructure, which helped give the economy a running start. Later, manufacturing centres began to develop outside these three cities thanks to investments in world-leading transport infrastructure and a multi-decade government commitment to regional development, as described in Chapter 3.

Economic development in Japan was tied to a number of changes in social policy. In the pre-war period, the Japanese government had actively sought to boost the country's population, banning birth control and advocating traditional notions of womanhood focused on domestic life. In the post-war decades, however, this stance was reversed, as the government became increasingly wary of the impact of high population growth on the nation's development prospects. Abortion was legalized in 1948, albeit with conditions, and a national campaign was launched encouraging family planning.[19,20] The result was a sharp fall in fertility rates from roughly five children per woman in 1930 to around two children per woman in 1960, driving a marked deceleration in population growth.[21]

Slowing population growth contributed to increasing labour shortages in the booming urban manufacturing centres, pushing up wages and helping to draw even greater numbers of inhabitants away from the countryside into cities. At the same time, agricultural productivity continued to improve on the back of mechanization. The result was that over the course of just a few decades Japan had successfully transitioned to an advanced, urbanized economy with a standard of living comparable to other rich countries.

For South Korea, the transition has been even more impressive. While richer than Spain today, in 1960 South Korea had an income per capita below that of Haiti, the poorest country in the Western hemisphere today.[22] The partitioning of the Korean peninsula at the end of Japanese occupation in 1945 effectively cut the South off from the abundant natural resources of the North.[23] In the years following the end of the Korean War, the country was destitute and heavily dependent on foreign aid for survival.[24] At the beginning of the 1960s, as the US planned an end to its reconstruction aid, the economic prospects for the country looked bleak.

In 1961, General Park Chung-hee seized power in a military coup, and set about emulating the export-led industrialization model that was proving so successful in Japan. Given the country's abundant cheap labour, the initial focus was on light industries such as textiles. Seoul, already possessing a critical mass of population and basic infrastructure, quickly emerged as the focal point for these industries, with industrial estates like Guro established by the government in the city to help spur the growth of a manufacturing economy.[25]

Trouble, however, soon began to emerge. While Seoul was experiencing robust growth in manufacturing employment, the population of the city was growing at a much faster rate, leading to mass urban poverty.[26] In 1970, the fertility rate in South Korea was still around five children per woman. With life expectancy fast improving, the result was very high rates of population growth. Agricultural yields, while improving, were not doing so quickly enough to feed the booming population in the countryside, leading large numbers of people to move to Seoul out of desperation.

In the 1970s, the government pivoted its development strategy in light of these challenges. The adoption of modern agricultural practices, including the breeding of improved cereal varieties and the increased use of chemical fertilizers, was accelerated. By 1980 rice production per hectare had caught up with Japan, helping to stem the rise of urban poverty resulting from rural flight. Moreover, the government decentralized production away from Seoul by encouraging the development of heavy industries such as steel, petrochemicals and shipbuilding in a number of smaller cities in the southeast of the country where flat land was plentiful and ports could be built on the East China Sea.[27] Seoul remained by far the largest city in South Korea, but the geographic widening of job opportunities helped improve access and ease excessive pressure on the capital. The transition to these more capital-intensive manufacturing industries had the added benefit of moving the economy away from a reliance on cheap labour as its source of competitive advantage in international trade, which became increasingly important as population growth finally began to slow. By 2000, South Korea had reached an urbanization rate above that of the United States and the major economies of Europe, and had cemented itself as a high-income country and a powerhouse in global manufacturing. Today, South Korea has the lowest fertility rate in the world, at 0.9 children per woman.

In neither country has the path to development been entirely smooth. In South Korea mass evictions were common. In both countries an overreliance on debt to fund rapid infrastructure development and capital accumulation set the scene for major financial crises in the 1990s. The impact of sharp declines in fertility is now also visible in the rapidly ageing populations of both countries. Yet much higher levels of prosperity have also been engineered over the space of just fifty years, a transformation anchored in the development of cities.

The experience of China is even more impressive. Over the past four decades, more than 800 million people have been lifted out of extreme poverty in China, which alone has reduced the number of desperately poor people in the world by 75 per cent.[28] At some point in the 2020s, China will probably pass the World Bank threshold for a high-income country, an incredible turnaround given that in 1950 it ranked in the bottom 20 poorest countries in the world.[29] As elsewhere, cities have been the engine room for this development. Never before have so many people moved in such a short time from the countryside to cities. By 2030 one in five city dwellers in the world will be Chinese.

The economic modernization of China began in earnest at the close of the 1970s under the leadership of Deng Xiaoping. First for reform was agriculture, where land was decollectivized and the 'Household Responsibility System' was introduced, shifting responsibility for profits and losses back to farmers. This set the foundations for a rapid acceleration in improvements in cereal yields, as farmers became incentivized to adopt more efficient practices. Next was industry. Rather than opening the economy to foreign influence all at once, the Chinese leadership established 'Special Economic Zones' in four coastal cities – Shenzhen, Shantou, Zhuhai and Xiamen – in which foreign multinationals were offered tax incentives and a more business-friendly operating environment in exchange for establishing export manufacturing facilities. The rapid success of these early experiments led to the designation of a raft of 'Open Coastal Cities' in 1984 that integrated China into global supply chains.

As in Japan and Korea, China's booming population presented a potential hurdle for its transition to a prosperous urban society. By the late 1970s fertility rates had already fallen to around three children per woman, thanks to a combination of concerted government media campaigns and the development of China's own version of the contraceptive pill.[30] However, with rapidly falling mortality rates, the population by 1980 had already hit one billion people, having nearly doubled since the Communist Party took power in 1949. Alarmed by the prospect of continued rapid population growth, the country's leadership in 1980 took the drastic step of introducing its 'One Child Policy' to curtail the country's birth rate.

While China's rapid economic development in its initial years was focused on the string of cities along its coastline, development soon spread to inland cities such as Chengdu, Wuhan and Chongqing, which were transformed into major centres of manufacturing activity. Together these cities absorbed a growing share of the rural population. The development of what is now the world's largest high-speed rail network has been an important factor in fuelling urban growth in China's inland regions, allowing them to connect with the country's coastal ports.

Today, China is in the process of consolidating its sprawling network of cities into 19 'supercity clusters'. The largest of these, the Pearl River Delta cluster, was described in the Introduction. The Yangtze River Delta cluster centred on Shanghai and the Beijing-Tianjin-Hebei cluster will be of a similar magnitude. Together, these 19 clusters are expected to account for 80 per cent of the country's GDP by 2030.[31] China's planners envisage two horizontal and three vertical 'corridors' of land and water linkages that will connect these urban clusters together.[32]

Avoiding diverging living standards between coastal and inland cities has been a perennial challenge for China's leaders. Historically, the focus was on building the industrial base of inland cities through subsidies and state investment in infrastructure. Recently, however, the process of convergence between coastal and inland cities has slowed as port cities like Shenzhen and Shanghai have developed into thriving innovation hubs. That has led the government to shift its focus towards turning inland cities into thriving hubs for emerging technologies too.[33] A new tranche of subsidies focuses on supporting technology companies in inland cities, and investments in enabling infrastructure like 5G have been ramped up. Inland cities have also been working hard to become more attractive to young knowledge workers by supporting nightlife, entertainment and leisure centres, and improving services such as healthcare and education.

The '*hukou*' system in China has long complicated efforts to reduce geographic disparities in living standards. Under this system, households are registered to their place of origin, and are unable to access basic entitlements like social security, education and healthcare elsewhere. If they relocate for work, they often end up in precarious employment without basic labour protections. That has resulted in significant hardship for rural migrants working in China's cities, and large and persistent levels of urban inequality. It has also resulted in the wrenching separation of families as working-age adults migrate to cities for work and children and elderly parents stay behind in the countryside. Recent moves to abolish *hukou* restrictions in cities with fewer than three million people are a step in the right direction, but this needs to be extended to all cities, enabling people to move with their dependants to wherever there are opportunities in China.

URBANIZED BUT POOR

The close link between urbanization and development in Japan, South Korea and China has not been universally replicated across the developing world. Consider the Democratic Republic of Congo. With a 46 per cent urbanization rate, the country has a higher portion of its population living in cities than Egypt, a country that is about six times richer.[34] In fact, when compared against the experience of rich countries during industrialization, most poor countries since 1950 have urbanized much faster than they have lifted incomes.[35]

In China, the authorities have exercised a powerful role in planning the growth of cities, whereas elsewhere the urbanization process tends to be more disorderly. In India, for example, cities have extended well beyond their administrative boundaries and growth tends to be largely unplanned, with people settling wherever possible on open land. Of the 30 cities in the world with the worst air pollution, 22 are in fast-growing India.[36]

Urbanization in the absence of development is associated with deep poverty in many cities across the developing world. A testament to this sad reality is the high share of urban populations in these countries that live in informal housing settlements, commonly but ungenerously referred to as slums. While London and New York had such settlements of their own in the late nineteenth century, their scale is far greater in poor countries today. In the Democratic Republic of Congo, roughly 80 per cent of the urban population lives in such areas.[37] In Mumbai, the population living in informal settlements is estimated at over seven million people, more than the population of Denmark.[38] Dharavi, which featured in the movie *Slumdog Millionaire*, is one such area. If you thought Manhattan was crowded with roughly 160 square feet (15 square metres) per person on a typical weekday, consider that there are just 30 square feet (fewer than 3 square metres) per person living in Dharavi.[39]

Living conditions in these informal settlements are typically very poor. Most residents lack a private toilet, and tend to rely on communal pits or even open spaces.[40] Many lack garbage collection services, which means rubbish may be left by the road or burned near the home. Reliance on traditional fuel sources such as kerosene, charcoal and candles for light, cooking and heating creates a fire risk that is compounded by population density and an absence of fire services. Access to clean water is patchy and tends to rely on communal taps shared by many families. The control exercised by monopolistic businesses or gangs over many informal settlements often compounds the problem, as residents are forced to pay prices for basic services well above fair market rates.

The internal migration driving the growth of these settlements reflects the grim reality that living conditions in rural areas are even worse. Lagos, the main city in Nigeria, has less than half the rate of extreme poverty as the countryside.[41] And while roughly 25 per cent of the residents of Lagos do not have access to clean drinking water, the same figure is an astonishing 70 per cent in rural areas.[42] The rural lifestyle in many poor countries is one on the edge of survival, characterized by frequent malnourishment and a lack of access to the jobs or education that might break the cycle of poverty. Many rural inhabitants in poor countries like the Democratic Republic of Congo have also been forced to flee to the cities for protection during civil wars, typically seeking refuge in already overcrowded informal settlements. As we will explore in more depth in Chapter 9, climate change is exacerbating this problem.

While many dwellers in informal settlements crowd into minivan taxis or walk along busy roads to get to their jobs in occupations such as cleaning or domestic service, many others lack regular employment and make ends meet in the informal economy, selling street food, collecting rubbish, repairing clothing and in other ways contributing to the bustling local economies. These irregular and unprotected jobs represent over 50 per cent of all urban employment in developing economies.[43] In cities like Kolkata in India, it is more than 80 per cent.[44] While the informal economy is a major source of livelihood for both men and women, the majority of these jobs are taken up by women.

Life in the informal economy is too often precarious. Sadly, some individuals are forced into prostitution or crime out of desperation. The favelas of Rio de Janeiro, for example, are home to warring drug gangs that play a major part in the city's high homicide rates. The policy of the government in the 1960s of relocating informal settlements to the periphery of the city, out of sight but also out of reach of many jobs, has left a lasting legacy.[45] The widely acclaimed 2002 film *City of God* brought to life the violent world of the inhabitants of one such favela.

The conventional 'modernization' theory of informal settlements suggests these places are a natural and transient phenomenon in the process of economic development. As incomes rise, makeshift houses are replaced by planned developments with access to roads, sewerage, water, electricity and other infrastructure and services. Yet there isn't much transience in the informal settlements of many poor countries today. In countries like Bangladesh they continue to grow, and even in moderately prosperous countries like Brazil their size has remained stubbornly consistent for decades. While some who live in these settlements manage to earn enough money to escape them, most do not, finding themselves stuck in place for decades, watching their children grow up in the same conditions they did. Indeed, these places can trap families in poverty for multiple generations, as the poor health and education of children hinders their prospects.[46]

As a result, some countries have made the eradication of informal settlements an explicit government priority. For example, in 2016 the government of Egypt began a five-year programme that successfully relocated roughly 850,000 people into subsidized social housing, and it has since set a further ambition to end all informal settlements by 2030.[47,48] The programme, which has involved vast amounts of demolition and redevelopment, also has created numerous challenges. While the new housing areas may be a step up in terms of living conditions and access to basic services such as health and education, many of those relocated have been cut off from the communities and support groups they lived with and the livelihoods they generated from the informal economy or nearby workplaces.[49] If relocated residents fail to find gainful employment, the risk is that these newly developed projects rapidly deteriorate into ghettos of concentrated poverty. The wrenching personal implications of these slum clearances is movingly described in Rohinton Mistry's book *A Fine Balance*.

An alternative approach has been to focus on upgrading existing informal settlements. One method for this that has gained popularity among some economists is to make property markets work more effectively. This school of thought, led by the Peruvian economist Hernando de Soto, starts with the observation that people in these places often lack the incentives or financial means to invest in improving their homes and neighbourhoods due to the absence of formal ownership rights and a high risk of eviction.[50] Granting land titles is seen as the solution. These would not only make it possible for those living there to invest in improving their homes, but also allow them to more easily take out loans for setting up small businesses or educating their children.

While seen as a magic bullet by some, the results of land titling programmes have often been underwhelming, suggesting they are far from the panacea hoped for.[51] There are no quick fixes to the challenge of informal settlements. Ensuring access to essential services – clean water, sanitation and electricity, alongside healthcare and education – is essential, though often difficult for governments with meagre finances. Even more important is creating the types of decent-paying jobs that can allow residents to escape from poverty. Doing that requires understanding and addressing the root causes of urbanization without development.

ROOTS OF THE PROBLEM

In Chapter 4, we described the terrible living conditions in cities during the early stages of the industrial revolution. Particularly problematic was the constant burden of infectious diseases in these overcrowded cities. Since then, rapid progress has been made in combating such diseases – a journey we explore in greater detail in the next chapter. Poor countries that urbanized later benefited immensely from the ground that was broken in countries such as Britain and the United States, with much lower mortality rates at comparable income levels.

For the avoidance of doubt, we feel passionately that fewer children dying and more people living into old age are causes for celebration. The tireless work of institutions and individuals to improve public health in poor countries is inspirational. Nevertheless, falling mortality rates have brought a dramatic expansion of population in poor countries. As infant mortality rates fall, women tend to have fewer children. The challenge is that declines in fertility tend to occur with a lagged effect. In many cases, that has triggered a desperate flight to cities as subsistence communities in the countryside struggle with many more mouths to feed.

In the major economies of East Asia, explosive population growth was brought down by the government policies we described earlier. But in India, for example, fertility rates were slower to transition downwards. While the birth rate has now dropped to roughly replacement level, the legacy of high fertility rates and falling infant mortality has left the country with a bulge of young people who will continue lifting population numbers over the coming decades. Poverty, compounded by climate change, is contributing to a rural exodus.

Urbanization itself can be an important factor in lowering fertility rates. Across developing countries, urban women have 1.5 fewer children on average than their rural counterparts.[52] Women in urban areas typically spend more of their reproductive years in education, are more likely to be employed in work outside the home, and are generally wealthier, which reduces the need to rely on children as a safety net and old-age pension. In rural areas, young children often support the family directly by tending to livestock or crops. While some children in urban areas are also forced by circumstances to continue working, the share of time they spend in school tends on average to be much higher than in rural areas, which weakens the economic rationale for large families. Awareness of and access to contraception and other family planning methods is also greater in urban areas.

Altogether, the result can be dramatic differences in fertility rates from one generation to the next in developing nations. Gautam Adani's parents were originally from a small town in rural Gujarat, but moved with their eight children to the major city of Ahmedabad to seek a better life for their family.[53,54] Gautam's father built up a successful textile business, which paid the way for the children's education, while his mother stayed at home to care for them. Adani, who would go on to become one of the richest men in India, married a qualified dentist, with whom he would have just two children of his own.

In a number of countries in Sub-Saharan Africa, however, fertility rates remain very high. For example, Nigeria currently has a fertility rate of about five children per woman and an infant mortality rate of just over 55 in every 1,000 births.[55] When India had the same infant mortality rate as Nigeria, its birth rate was around three.[56] Low rates of contraceptive use in a number of countries in Sub-Saharan Africa are one factor. Also important are restrictive views about the role of women in society, particularly in the Sahel area just below the Sahara Desert. These repressive beliefs lock many women out of employment, hold back the education of girls, and reinforce the dominant role of men in the family, all of which perpetuate high fertility rates. While fertility rates have fallen substantially in a number of Sub-Saharan countries such as Kenya and Ethiopia, reductions have been slower in others like Nigeria and the Democratic Republic of Congo.[57]

The delayed modernization of agriculture in many poor countries has also been a contributing factor in the accelerated pace of urbanization. The link between agricultural productivity and urbanization is nuanced. Rapid growth in agricultural output above and beyond population growth can be a driver of urbanization, as a rising share of the population is freed from the need to work the land. That is the pattern that generally prevailed in countries like the United States throughout the process of industrialization. But in a developing country like Uganda that is experiencing rapid population growth, it is also possible for *too little* agricultural production to drive urbanization, as people seek to avoid starvation by moving to urban areas where they may be able to find a job. Food supply and trade is far more globalized today, which means that urbanization rates are no longer constrained by a country's own agricultural surplus.

There is considerable debate around why agricultural productivity has been slow to improve in many poor countries. One explanation lies in the development strategy known as 'import substitution industrialization' adopted in the decades immediately following the Second World War in a number of developing countries – most enthusiastically in Latin America, but also in South Asia and Sub-Saharan Africa. In contrast to the export-oriented development model followed in East Asia, this approach focused on replacing the import of foreign manufactured goods with domestically produced ones by shielding local industry from foreign competition. The challenge, to simplify, was that the small size of domestic markets in these countries meant that protected industries often lacked the scale necessary to be viable.[58] Significant amounts of public funding were needed to keep these industries afloat, leaving little for sorely needed agricultural investments such as modern irrigation systems. Often it also involved indirectly subsidizing the wages of manufacturing workers by capping agricultural prices below market rates, which reduced the potential for farmers to be profitable and to invest.[59]

The import substitution model has since fallen out of favour around the world, and many poor countries have benefited from the modernization of their agricultural systems in recent decades. In Sub-Saharan Africa, however, agricultural yields remain well below the rest of the world. In Nigeria, cereal production per hectare is a quarter of what it is in China.[60] Part of this is due to poor soils and, increasingly, to climate change, droughts and extreme weather, an issue we return to in more detail in Chapter 9. But this is only a partial explanation. Agriculture in many places is stuck in a different era, with limited use of inorganic fertilizers and other chemical products, a lack of modern machinery like tractors, and an absence of improved seeds. The interaction between low rates of improvement in agricultural productivity and high rates of population growth goes a long way to explaining the continued rapid rural-to-urban migration and the growth of informal settlements in many developing countries.

So far we have focused on what causes the premature growth of cities in many poor countries. Equally important in urbanization without development is a stalled process of industrialization. Manufacturing jobs, by offering mass employment to the urban poor, have been an essential stepping-stone in the path to prosperity in those countries that have escaped poverty.

There is no single explanation for why poor countries fail to industrialize, and providing a full account is not the focus of this book. But a few salient points are worth noting. The first is that the business environment matters. The experience of the three East Asian economies described earlier illustrates that successful rapid industrialization typically involves an active role for government in shaping the development process. But excessive meddling, particularly when combined with a corrupt and ineffective government, can prevent markets from effectively allocating capital and people to the most productive enterprises, leading the process to grind to a halt.

Civil war has also been an important barrier to industrialization in many poor countries, especially for those that were arbitrarily carved up by colonial overlords, setting the scene for repeated clashes between ethnic groups that lack a unifying national identity. Civil conflict not only leads to the destruction of the infrastructure necessary for industrialization, but also undermines domestic and foreign investment. Since 1960, the average civil war has lasted eight years, with incomes taking 14 years just to recover to their pre-conflict levels, a significant setback in the process of development.[61] Joblessness and weak state institutions in countries impacted by civil war can set the stage for repeated cycles of internal violence that prevent industrialization ever taking off.

An overreliance on natural resource exports can also obstruct industrialization. While discovering oil or gold might sound like a stroke of good luck for a poor nation, economists are more wary, and talk of the 'resource curse'. When a country sells these resources abroad, foreign buyers need to purchase its currency in order to fulfil the transaction, which pushes up the exchange rate of the exporting country. The problem is that a higher exchange rate subsequently makes it more expensive for foreign buyers to purchase anything else from that country, including its manufactured products, which puts a brake on industrialization. It also makes it more expensive for foreigners to visit and stifles tourism. A heavy reliance on natural resources can also contribute indirectly to excessive urbanization. The capital-intensive nature of modern resource extraction means most of the value it generates flows into the hands of a small number of wealthy elites, the majority of whom reside in the major cities of their countries.[62] This helps to explain why places like the Democratic Republic of Congo, with its vast deposits of minerals, are highly urbanized without being meaningfully industrialized. This pattern is evident in many resource-rich countries in Africa and the Middle East, as well as in Latin America and a number of economies in Southeast Asia. Catering to the needs of elites may offer a source of employment for the urban poor, but it is not a path to broad-based prosperity. This economic model is associated not only with high levels of inequality, but with high volatility too, as national living standards rise and fall with global commodity prices.

Successfully navigating the pitfalls of rapid and poorly managed urbanization combined with insufficient industrialization is a tall order, and explains why the experience of the three East Asian economies described earlier is the exception rather than the norm. Looking ahead, the question is whether it will be possible for other poor countries to follow the path carved by those trailblazers, or whether the nature of technological change will irreversibly alter the process of economic development.

A BROKEN LADDER?

The stylized ladder of economic development starts with labour-intensive activities when workers are abundant but physical and human capital is scarce, transitioning to capital-intensive activities as investments accumulate and infrastructure develops, and finally moving on to knowledge-intensive activities – such as the development of new technologies and the management of global corporations – as the workforce becomes both educated and expensive. We now see this final step at work in East Asia. Japan is the global leader in robot production, while South Korea has the second highest rate of R&D spending as a share of GDP in the world, second only to Israel.[63] China, heavily dependent on cheap labour only a few decades ago, is increasingly emerging as a technological powerhouse in its own right.

However, as noted in Chapter 3, the increasing automation of manufacturing in rich countries has made production far less labour-intensive than it once was. As technology increasingly reduces the need for abundant unskilled labour, it is unclear what the source of competitive advantage will be for countries still in the early stages of industrialization. Bangladesh has established itself as a major manufacturer in the labour-intensive garment industry thanks to its abundant and cheap workforce.[64] But robots capable of sewing could soon pull the rug out from underneath Bangladesh's manufacturing sector.[65] For poor economies that have not yet started on the journey, the prospects look even more bleak. With many rich countries actively looking to bring a greater share of manufacturing back home in a bid to keep voters happy and reduce reliance on global supply chains, the challenge is likely to grow.

In the absence of the traditional path to industrialization, what other avenues for development are available? Many poor countries blessed with attractive scenery or a warm climate will continue to benefit from tourists. But the jobs this sector creates tend to be relatively low paid, and many poor countries do not have the beaches of Thailand or the wildlife of Tanzania. Another option is to provide cheap call centres or clerical services such as payroll. Many of these highly routinized activities, however, are already being automated, with recent developments in generative artificial intelligence increasing the risk of replacement.

A new consensus on migration could be one part of the solution. While many of the poorest countries of the world will continue to experience rapid population growth in the decades ahead, many rich countries – and some middle-income countries like China – are experiencing rapid ageing and will see their populations contract in the coming decades. Japan is a harbinger of this trend. Its population has been declining since 2011 and by 2050 may have shrunk by 30 million, almost a quarter. In Japan, as in many other countries with rapid ageing and declining fertility, a growing number of the retired elderly will need to be supported by a shrinking working-age population. Low-skill jobs that are difficult to automate – in areas like home care or hospitality – will become even more difficult to fill, leading to chronic job shortages. Allowing this to occur while billions in the developing world are desperately seeking employment would be absurd. Across rich countries, immigrants have been demonized in recent years for supposedly stealing good jobs, but they are not to blame for the hollowing out of the middle class. Political courage is needed to find sensible solutions, which will almost certainly require more immigration. We are not advocating simply opening borders. Migrants need to be documented, not just to ensure they pay taxes but also to protect them from exploitation. And robust controls need to be complemented by efforts to ensure that migrants integrate effectively into society, which is essential for maintaining public buy-in. With these conditions in place, increased migration can be a win-win, helping rich countries to overcome their demographic challenges while simultaneously reducing global poverty by providing a lifeline to millions of people around the world desperately seeking work. Equally, there are more than five times as many people in poor countries as there are in rich countries, which means that the lion's share of the work in tackling global poverty will need to occur within developing countries.

Some hold out hope that poor countries can 'leapfrog' to the top of the development ladder by focusing on turning their cities into hubs for knowledge work and innovation. A promising example of this is Bangalore in India. The city has combined a long history as a centre for IT outsourcing with strong local universities to build a skilled talent ecosystem that is able to go head to head with many developed world knowledge hubs. Today, the city hosts offshore R&D centres for Amazon, Microsoft, Facebook and Google, and generates $130 billion in exports annually.[66] By matching its talented engineers with a pool of venture capital investment equal to that of Berlin, the city has also spawned a number of domestic digital champions, from e-commerce firm Flipkart to ride-hailing firm Ola.[67]

Other poor countries are seeking to pull off similar feats. For example, the Yaba district of Lagos – referred to by locals as 'Yabacon Valley' – has emerged as a focal point for digital talent in Africa. But like most such clusters, it remains small in scale. Changing that will require massive investment in higher education, supported by reliable communication infrastructure and a lot more venture capital.

The question of how today's poor countries will rise to prosperity is both difficult and urgent. Without an answer, the risk is that the cities in many developing countries will cease being a mechanism for eradicating global poverty and devolve into areas of sprawling deprivation. This would be not only a tragedy for the inhabitants of the developing world, but also an existential risk for humankind, as we will show in the next chapter.

8

The Spectre of Disease

In 1348, Venice was the most prosperous city in all of Europe, and one of its largest. As the gateway for trade into the continent, the city was filled with merchants from across the known world, generating a distinctively cosmopolitan atmosphere. It was also a magnet for rural migration, with farmers seeking refuge from the famine of the preceding years.

1348 was also the year the Black Death arrived in Venice on a merchant ship. By 1352, the plague had claimed the lives of around 60 per cent of the city's inhabitants.[1] High population density provided the ideal conditions for its rapid transmission between inhabitants, while the city's reliance on its global connections made it difficult to limit the ongoing influx of exposed travellers. It took Venice more than 30 years to institute its first quarantine – derived from the Italian *quaranta giorni*, meaning 40 days – under which the city-state required all ships arriving from infected ports to sit at anchor for 40 days before landing.[2]

Once the Black Death – or bubonic plague – was established in the trade hubs of the Mediterranean, it spread like wildfire across Europe, carried by travellers passing between cities. Ultimately, it is estimated to have killed 40 per cent of Europe's population.[3] The Italian writer Giovanni Boccaccio, who survived the plague in Florence, provided a moving account of the horror, as 'dead bodies filled every corner ... brother abandoned brother ... [and] fathers and mothers refused to see and tend their children, as if they had not been theirs'.[4]

Pandemics have long been the great wildcards of history. The Antonine Plague that crippled Rome during the second century CE is considered by some historians to have been a major factor in the empire's subsequent decline.[5] Similarly, the Justinian Plague that tore through Constantinople during the sixth century CE ended the Byzantine emperor's dream of reconquering Rome and once again uniting the Eastern and Western halves of the old empire.[6] The appalling loss of life in Europe during the Black Death was a major factor in the collapse of the feudal system, and helped to undermine the perceived legitimacy of the Church.[7] The rapid spread of infectious diseases like smallpox that Europeans carried to the Americas decimated local civilizations and made both the northern and southern continents more easily conquerable by the invaders.[8] Without these events, the world would look very different today.

In Chapter 2, we explored how the transition from isolated rural communities to globally connected cities has been pivotal in human progress. But it has not come without cost. Small communities with limited connections to one another provide little opportunity for infectious diseases to spread. Cities, on the other hand, with their high population densities and transport connections, provide the ideal conditions for diseases to extend their reach, evolve and thrive.

From the dawn of civilization until only recently, the biggest check on human population was not famine or war but the continual threat of infectious disease.[9] New diseases like the bubonic plague would periodically burst onto the scene and rapidly disseminate via the transport routes connecting cities, triggering uncontrolled pandemics that would claim many lives in a short period of time. In some cases, they would eventually run out of hosts and die back. In other cases, like smallpox, they would establish a permanent foothold in the human population, joining forces with other endemic diseases such as malaria and measles to provide a constant source of misery and death.

The twentieth century saw remarkable progress in combating contagions. Modern urban sanitation reduced the transmission of waterborne diseases like cholera or typhoid. Vaccines made possible the eradication of smallpox and the taming of other viral diseases such as polio and measles. Antibiotics have taken the sting out of the bacteria that cause contagious diseases like tuberculosis and even the bubonic plague. In 1987, on the back of these triumphs, the US Surgeon General William Stewart boldly declared: 'The war against diseases has been won.'[10] Even in poor countries, where rates of death from preventable infectious diseases are still much higher, progress has been enormous. Infectious diseases seemed poised for defeat.

The devastating toll of the Covid-19 pandemic has shattered this naiveté. That global pandemic has led to the loss of over 20 million lives, a similar death toll to the First World War.[11] Most of these deaths have occurred in poor countries, though rich countries have certainly not been spared, with over one million deaths in the US alone. Many of our readers will have lost loved ones, with the burden of mortality having fallen disproportionately on the elderly, medics and essential workers who were most exposed to the pandemic, particularly in the first year before vaccines became available. Others will be suffering from 'Long Covid', whose lasting effects we are only beginning to understand.

The impact of Covid-19 has stretched far beyond lives lost.[12] Lockdowns, which were essential to slow the spread of the disease and prevent the collapse of health systems, triggered the most severe global recession since the Great Depression. In rich countries, governments stepped in with economic stimulus packages that dwarfed those put in place during the global financial crisis. These softened the economic blow of lockdowns significantly, but will leave a long legacy of government debt at a time of already-strained public finances and sluggish economic growth.[13] In poor countries, governments simply did not have the means to conjure up that kind of support, leaving many households confronting the difficult choice of stopping work and struggling to put food on the table, or continuing to work at risk of infection. To make matters worse, travel bans decimated the jobs in tourism, hospitality and entertainment that provide a lifeline to many, and especially so in poor countries. Around the world, nearly 100 million more people were living in extreme poverty in 2020 than there would have been without the pandemic.[14] A slower recovery in poor countries unable to jolt their economies back to life with government stimulus means the devastating impact on global poverty will be long lasting.[15] The food and energy crisis precipitated by the Russian invasion of Ukraine has further exacerbated dire poverty. There are now 200 million more people around the world facing acute food insecurity than pre-pandemic.[16]

Beyond its economic impact, the pandemic has also taken its toll in other ways. Rates of anxiety and depression in the US increased roughly six-fold in 2020, with younger adults hardest hit.[17] Most people do not cope well when cut off from social interaction. The fact that many young people were forced to delay major life milestones such as entering the workforce or getting married also contributed to the misery.

For the generation of children currently in school, the after-effects of the pandemic will be felt for decades. In the two years following the onset of the pandemic, schools around the world shut their doors for an average of 20 weeks, and remained only partially open for roughly the same amount of time.[18] In rich countries, children from privileged backgrounds – with all the right technology and a supportive home environment – fared better than their less privileged counterparts. In Britain, learning losses were 50 per cent greater in secondary schools in disadvantaged neighbourhoods.[19] The impact of lost time in the classroom was by far the greatest in poor countries. In the Philippines, children received no schooling of any kind for seven months.[20] Across poor countries, the share of ten-year-old children unable to read a simple story has risen from 57 per cent to 70 per cent as a result of the pandemic. Globally, the World Bank has estimated a price tag of over $20 trillion dollars in lost lifetime earnings for this generation of schoolchildren.[21]

To allow the memory of this disaster to fade, dismissing it as a once-in-a-hundred-year event, would be a terrible mistake. Humanity is at far greater risk from catastrophic pandemics than many recognize. The history of humanity's relationship with infectious diseases can be summarized as a series of epidemiological transitions. The first was the rise in pathogens that accompanied the birth of civilization, as the density and interconnectedness of human populations increased. The second was the subsequent fall in infectious diseases over the past two centuries thanks to rapid advances in our understanding of the underlying mechanisms behind these pathogens. Some epidemiologists believe we are now entering a third transition, in which the emergence of contagious diseases is accelerating towards a speed we can no longer control.[22] The coronavirus pandemic, as destructive as it has been, may prove merely a precursor for what is still to come.

Ian's 2014 book *The Butterfly Defect* explored how the world was becoming more vulnerable to risks like pandemics. Cities, by bringing the world together, have played a central role in this. About 250 people can fit into a single carriage on the New York subway, greater than the size of a typical medieval village.[23] In 2019, prior to the pandemic, the global aviation industry completed 4.5 billion passenger journeys, nearly all of which departed and arrived in cities.[24] Only through cities can an outbreak of a novel disease in central China cascade in a matter of weeks into a global pandemic.

THE ORIGINS OF INFECTIOUS DISEASE

When our ancient ancestors transitioned from hunting and gathering to agriculture, they forever changed our relationship with the world of infectious diseases. Rising population density made it easier for microorganisms to establish themselves in multiple hosts. Greater exposure to domesticated livestock and pets increased opportunities for infectious diseases to make the jump from animals to humans. And more human waste and refuse in sedentary settlements attracted the insects and rodents that help transmit disease.

The small scale and relative isolation of agricultural settlements kept a limit on the potential for large and sustained outbreaks of pathogens. If a disease were to make the transition from a domesticated animal to a human in a small village, the chances were that it would not spread beyond that community. And for an infectious microorganism to become endemic in a population, it needs a continual chain of new hosts to infect, something an isolated village simply does not provide. It would take the emergence of the city, with its sizeable population and greater interconnectedness, to give pathogens the opportunity to spread and mutate, putting societies at their mercy.

Greater exposure to infectious diseases has meant that inhabitants of cities, despite on average being richer, have throughout history tended to experience higher mortality rates than their rural counterparts.[25] Indeed, the increased danger of disease has historically been one of the limiting factors on the size of cities. At a time when the physical extent of a city was limited by walking distance, any expansion in population would mean greater density and therefore increase the likelihood of a higher mortality rate from infection.

This is precisely what occurred during the early industrial revolution. Fast-growing cities saw a surge in a number of infectious diseases on the back of rising populations. Sanitation systems ill-equipped to cope with rapidly expanding numbers of inhabitants led to frequent outbreaks of diseases like cholera and typhoid, transmitted through tainted water supplies, while increasingly crowded living conditions increased the likelihood of person-to-person transmission of diseases such as influenza and tuberculosis. As a result, the cities of the early industrial revolution became death-traps for their inhabitants, expanding only thanks to rapid inward migration of impoverished rural populations.

TURNING THE TIDE

Control over infectious diseases is one of the great success stories of modernity. There is no grief quite like that of a parent burying their child. Fortunately, only four in every 1,000 children in Britain today die before the age of five. In 1800, however, that figure was a heart-wrenching three in every ten, most of which was due to infectious diseases.[26]

To understand how so much progress was made in such little time, we need to reflect on how far our understanding of medical science has evolved in recent centuries. It is now widely understood that microorganisms like bacteria, viruses and parasites are the cause of infectious diseases. However, until the latter decades of the nineteenth century the prevailing wisdom was that bad air, or 'miasma', was to blame, an idea hailing back to Hippocrates of Ancient Greece.

The British epidemiologist John Snow was one of a small number of pioneering thinkers who found the theory unconvincing. In 1854, Snow created a map of where cholera cases were being contracted in the West End, and noticed a distinctive cluster around a single water pump in Broad Street, Soho. Snow realized that the cause of the outbreak was a tainted water supply, as the faecal discharge of cholera victims in the area was seeping back into the well due to the absence of an effective sewerage system. Snow convinced the local authority to remove the handle from the well, but he failed to persuade a subsequent parliamentary inquiry that the cause of regular cholera outbreaks was a dirty water supply rather than noxious air.[27]

At the time of John Snow's investigation, the most influential figure in sanitation in Britain was a social reformer named Edwin Chadwick. In 1842, Chadwick published a *Report on the Sanitary Condition of the Labouring Population of Great Britain*, in which he argued forcefully for cleaning up London to reduce the instance of disease. An ardent believer in miasma theory, however, Chadwick's solution was to get the supposedly noxious air away from houses by closing the city's cesspits and redirecting human waste through a network of pipes into the Thames River. With that river providing much of the city's supply of drinking water, Chadwick's solution only made the problem worse.[28]

It was not until the episode of the 'Great Stink' in 1858, mentioned in Chapter 4, that parliament decided to take decisive action to clean up the river itself. A major programme of upgrading London's sewerage system was initiated, with waste being pumped downriver of the city's water intake and flushed out to sea, freeing the city from its periodic cholera outbreaks.[29]

It would ultimately take the diligent scientific efforts of Louis Pasteur and Robert Koch for the germ theory of disease to finally supplant the miasma theory. Their discoveries led not only to efforts by cities around the world to clean up their water supplies, but also to the adoption of basic hygiene practices we now take for granted – like washing hands or cleaning medical implements after use – that play a crucial role in preventing disease transmission.

We also owe the wide range of vaccines we now have available to the pioneering thinkers behind the germ theory of disease. Vaccines have a long history. Inoculation against smallpox became commonplace in Europe early in the nineteenth century after an English country doctor named Edward Jenner discovered that infecting a patient with the far less dangerous cowpox conferred immunity against smallpox. It was not, however, until the late nineteenth century that Louis Pasteur was able to apply the insights of germ theory to create a more systematic approach for weakening bacteria and viruses to make them suitable for inoculations against a wide variety of diseases.

Likewise, Alexander Fleming's discovery of antibiotics would never have been possible without germ theory. Without antibiotics, the bubonic plague would still be a very real risk to the global population, while an infection of cholera, typhus or tuberculosis would prove far more lethal than it does today.

Advances in our understanding of the root causes of infectious disease allowed us to turn the tide against pathogens, and helped to remove the mortality penalty of urban life. That in turn has led to a rapid expansion in the global population as well as the share of people living in cities, setting the scene for a resurgence in infectious diseases. Many battles against diseases have resulted in resounding successes, but the war against pathogens is not over.

A NEW AGE OF PLAGUES?

Decades before Covid-19, warning signs of a rise in infectious diseases had begun flashing. HIV in the 1980s, Bird Flu in the 1990s, and SARS in the 2000s were all dangerous diseases that had previously not been observed in humans. In fact, since the middle of the twentieth century, we have seen not just a rise in the frequency of infectious disease outbreaks, but a proliferation in the number of distinct pathogens affecting humans.[30,31]

Three-quarters of these new infectious diseases are what epidemiologists refer to as 'zoonoses', which means they have been transmitted from animals to humans.[32] As noted earlier in the chapter, the domestication of animals that began with agriculture has made it possible for various diseases like influenza to make the leap to humans. That process has accelerated over the past century as the total volume of livestock globally has risen sharply to cater to a much larger and much richer human population. Efforts to produce meat cost-effectively have led to far more animals being crowded together in confined indoor spaces. In the European Union, regulation permits a density of roughly 16 chickens per square metre. Elsewhere in the world where regulations are less stringent, particularly in poor countries, densities can be even higher. Such conditions facilitate the rapid transmission of infectious diseases among livestock, while the increased live transportation of animals over long distances makes outbreaks difficult to control.[33]

The disruption of wild habitats by the global population's expanding physical footprint has also contributed to the rising number of zoonotic diseases being transmitted to humans. In 1998, the fatal Nipah virus first appeared in Malaysia. While the disease spread to humans through infected pigs, it originally passed to those pigs via fruit bats that had been displaced by deforestation.[34] Such transmissions will continue to occur so long as we keep clearing 10 million hectares of forest around the world every year.[35] Most of this occurs in developing countries, with countries like Indonesia, Brazil and Malaysia deriving a significant share of their national income from export crops like soya, coffee or palm oil.

The consumption of 'bushmeat' – animals such as apes, rats, bats and pangolins – is another factor in the growth of zoonotic diseases. While a small amount of bushmeat is eaten as a delicacy, most consumption occurs due to the absence of alternatives. HIV and Ebola are likely to have made their way to humans through this practice.[36] It is estimated that as much as 5 million tonnes of wild animals are caught and eaten every year in the Congo basin alone.[37]

Most major new zoonotic diseases in living memory have emerged in the developing world, and are likely to continue doing so for the reasons mentioned above. Large and tightly packed populations in the cities of poor countries accelerate the subsequent spread of infectious diseases within and beyond national borders. The role of a single live-animal market in Wuhan as the likely epicentre of the Covid-19 pandemic clearly illustrates this point.[38] So does the first recorded outbreak of HIV in Kinshasa, the capital city of the Democratic Republic of Congo.[39]

In the decades ahead, humanity will continue to be linked ever more tightly together in cities, and these cities will continue to act as catalysing agents for the new diseases emerging as a result of our expanding ecological footprint. As a result, a global pandemic such as Covid-19 is unlikely to remain a once-in-a-hundred-year, or even once-in-a-generation event. Indeed, one study has put the likelihood of a similarly deadly pandemic occurring in the next 25 years at roughly 50 per cent.[40] The prospect of an even deadlier virus emerging remains very real. For example, the paramyxovirus family contains pathogens such as measles that are extremely contagious as well as ones like the Nipah virus that exhibit high mortality rates.[41] Some epidemiologists therefore worry that this family could produce a virus that combines both traits.[42] If not contained, the death toll from such a disease would be unimaginable.

Opposition to vaccines is also raising our risk from infectious diseases. In 1998, Andrew Wakefield, then a doctor in London, published an article in the *Lancet* medical journal suggesting a link between the MMR (measles, mumps and rubella) vaccine and autism in children. Since then, numerous studies have tested and refuted any connection, the original article has been retracted, and its author has been disbarred for professional misconduct.[43] Nonetheless, Wakefield's research has helped fuel the rise of an anti-vaccination movement in many countries. While the resulting spike in the number of measles cases has been modest, with the majority occurring in unvaccinated children, the bigger issue is the loss of faith in this critical tool of disease prevention. Vaccination is essential in maintaining herd immunity to infectious diseases across generations, without which periodic outbreaks are bound to occur. By giving viruses a chance to spread, those outbreaks raise the likelihood of new variants emerging that may be partially or fully resistant to available vaccines. While measles is a virus with relatively limited capacity to mutate, many others are not.[44]

Another concern is growing antibiotic resistance. Penicillin was discovered in 1929 and began being used systematically in humans in 1941. By 1942, penicillin-resistant strands of the *staphylococcus* bacteria had already started to emerge.[45] In recent decades, antibiotics have become increasingly ineffective due to overuse.[46] Some of this is down to over-prescription by doctors, but most of it has to do with the use of antibiotics in livestock and fish farming, not just for prophylactic purposes, but also for the coincidental side effect of increased animal growth.[47] Already, antibiotic-resistant pathogens are claiming the lives of over a million people every year and limiting our ability to treat endemic bacterial diseases.[48] There is a concerning possibility that a highly infectious bacteria such as *Yersinia pestis* – which causes bubonic plague – could emerge in the coming years with resistance to antibiotics. Indeed, a strain of *Yersinia pestis* resistant to eight different antibiotics was identified in Madagascar in 2007.[49]

Lastly, there is the increased risk that advances in biotechnology could be harnessed by malicious actors. Gene-sequencing machines have become both more advanced and more accessible in recent decades, while technologies such as CRISPR are making it easier to re-engineer organisms in targeted ways.[50,51] These scientific advances hold immense promise. The speed with which tests and vaccines were developed for Covid-19 would not have been possible without recent advances in genomic sequencing. And they may eventually make it possible to eradicate diseases like malaria and dengue fever for good by making mosquitoes resistant to the parasites responsible for them.[52] At the same time, these technologies open up possibilities for weaponizing diseases to maximize loss of human life. Global monitoring efforts make it extremely difficult for a terrorist organization to acquire the enriched uranium needed for a nuclear weapon, but far less stringent controls exist on the laboratory equipment needed to produce a biological weapon. This greatly raises the risks of laboratories located in major cities being the source of deliberately or accidentally created pandemics in the future.

As outbreaks of infectious disease emerge with increasing speed and severity, cities will play a decisive role in either aiding them along or stopping them in their tracks. Covid-19 offers us an opportunity to better understand that dynamic, and ensure we are prepared for what lies in store.

CITIES AND COVID-19

London's Heathrow and New York's JFK both offer connecting flights to over 80 countries and bring in millions of passengers from abroad every year. In both cases, they are one of multiple international airports operating in these cities. It is therefore not surprising that both cities were hit early by the Covid-19 pandemic, and it is almost certain that the initial outbreaks they experienced played a decisive role in the pandemic's subsequent spread both at home and abroad.

In addition to being more globally connected, bigger populations in major cities means faster transmission. That is why, across the US, larger cities experienced higher Covid-19 infection rates.[53] At the same time, as described in Chapter 3, major cities tend to be richer, which means they also tend to have healthcare systems better resourced to cope with a crisis. Richer people are also less likely to have pre-existing health issues such as obesity that contribute to higher mortality rates.[54] This helps explain why mortality rates from Covid-19 infections were 20 per cent higher in Detroit than in New York City.[55] Fewer hospital beds, on top of an older and poorer population, is also why mortality rates in the US from Covid-19 were higher in rural areas than in cities, a surprising reversal of the historical pattern.[56]

The Covid-19 pandemic also illustrated an important point about the unequal impact of infectious diseases within cities. In New York, Queens is only a third as densely populated as Manhattan, but has had 50 per cent more pandemic fatalities on a per capita basis.[57] In cities like London and Paris, the story is much the same. Residents of poorer neighbourhoods in these cities are far more likely to work in service occupations that require in-person interactions. In the United Kingdom, it is estimated that fewer than one in ten workers in the bottom half of the income distribution can perform their jobs remotely.[58] And residents of poorer neighbourhoods not only experienced higher rates of infection, but also a higher likelihood of dying from those infections, due to increased prevalence of pre-existing health issues. The disproportionate burden of infectious disease on the poor inhabitants of cities has long been a feature of pandemics. Skeletons from the time of the Black Death in London show that people with signs of disrupted enamel growth, an indicator of malnourishment, were disproportionately more likely to die from the plague.[59]

What of poor countries? Accurate comparisons are challenging given less testing, but based on comparisons to pre-pandemic death rates it is likely that as many as 90 per cent of all fatalities from Covid-19 have occurred in poor countries.[60] India and South Africa have had twice as many excess deaths per capita during the pandemic as Britain and three times as many as Germany.[61] Delayed access to vaccines has been a big part of this, but living conditions in the cities of the developing world have also contributed. Crowded living conditions in which many people share rooms makes isolation impossible, greatly increasing the potential for these places to become hotspots for transmission. The share of urban inhabitants in poor countries who are in jobs that could be performed remotely is also much lower than in rich countries, and the need to use crowded public transport to get to work further increased the risks of infection. In the absence of financial support from their governments, many had no choice but to continue working despite the risk of infection in order to keep food on the table. At the same time, under-resourced health systems struggled to cope with high rates of infection, which increased the likelihood that an infection could turn fatal.

Covid-19 also proved that what happens in the crowded cities of the developing world is of global concern. The more hosts an infectious microorganism passes through, the greater the likelihood that a new variant emerges that may be not only more contagious but also more resistant to vaccines or other treatments. It is not a coincidence that India and South Africa, two developing countries that were very severely impacted by Covid-19, both produced major variants that subsequently spread across the globe.

PREPARING FOR THE FUTURE

It is time to start taking the global risk of infectious diseases far more seriously. The budget of the World Health Organization is less than that of many large hospitals in the US or Europe.[62] It is in desperate need of more resources, new capabilities and a stronger mandate to detect outbreaks and coordinate international responses. Bill Gates has suggested a price tag of roughly $1 billion for a new dedicated unit of epidemiologists, geneticists and rapid-response workers within the WHO to stop future outbreaks before they become pandemics.[63] With the total economic cost of Covid-19 at more than $20 trillion, that seems like a bargain, amounting to a fraction of one day of the cost of the pandemic. The size of investment required also pales in comparison against what governments spend on national defence every year.

Greater global coordination is also needed to create safer food systems that minimize the chance of new pathogens emerging in the first place. Reducing excessive livestock densities, minimizing the live transport of animals, stopping unnecessary antibiotic use and discouraging the consumption of bushmeat all need to be part of this. Slowing and ultimately ending deforestation is also essential. The burden of such initiatives will fall disproportionately on poor countries, many of whose citizens rely on these practices to survive. If rich countries wish to reduce the risk of future pandemics, they will need to be willing to help poor countries pay for these changes.

Preventing outbreaks from spiralling out of control also requires transforming cities from pandemic catalysts to pandemic choke points. Lockdowns are a vital tool for reining in uncontrolled transmission, but experience with Covid-19 shows that they cause immense hardship and economic damage. Contact-tracing technology, which proved woefully inadequate during Covid-19 in all but a few countries, needs to be stepped up in advance of the next pandemic. Fears of invasive surveillance can be mitigated by making it transparent what data will be collected, under what circumstances, and how that data will be stored and used. Such improvements to transparency are necessary anyway, as cities gather ever more data on their citizens for uses ranging from policing to traffic management. Enhanced contact tracing needs to be combined with ubiquitous rapid testing that can be ramped up quickly in response to outbreaks. More tests and better tracing will allow cities to use shorter lockdowns, less often. These capabilities should be regularly pressure-tested through simulations, and channels should be established for sharing best practices around the world.

Cities also need to become more aware of the impact of inequality on the course of pandemics. For the foreseeable future, a large share of urban dwellers will work in jobs that require in-person interaction, with the lion's share of these jobs being filled by those on lower incomes. Some of these workers, including those in healthcare or stocking supermarket shelves, must continue working even during a pandemic. Cities need to build stockpiles of PPE that can be provided to these workers, guarantee access to regular testing and ensure they are first in line for vaccines. Those working in occupations that are not essential but cannot be done from home need to be supported financially during lockdowns. We should take the time in advance of future outbreaks to design mechanisms for distributing that money in a way that minimizes the potential for fraud and abuse while maximizing the speed of implementation. More broadly, the geographic coverage of healthcare needs to be reviewed to ensure that those living in poorer cities and poorer neighbourhoods have adequate access to medical care and hospital beds.

Finally, the interconnected nature of our world today means that living conditions in the cities of poor countries cannot be ignored. An outbreak in Kinshasa or Dhaka can rapidly evolve into a global threat. Improving health systems in these countries needs to be a global priority. That will not just ensure they are in a better position to handle future pandemics; it will also improve their ability to manage endemic diseases such as malaria that continue to claim far too many lives. At the same time, tackling the underlying lack of development that leads to overcrowding, unsanitary living conditions and the uncontrolled spread of disease in these cities should be viewed not as an act of charity, but as an essential link in the chain of defence against pandemics that know no borders.

Earlier in this chapter, we noted how pandemics have been the cause of a number of major inflection points in history. Equally, the fundamental forces that draw people towards cities have proven remarkably resilient in the face of the onslaught of infectious disease. The Black Death with which we started this chapter decimated the population of European cities in the middle of the fourteenth century. But by the start of the fifteenth century, Europe's urban population had more or less recovered to its pre-plague levels, with large numbers of people migrating from the countryside to repopulate cities.[64] But what if the original condition that made the birth of cities possible – a predictable climate favourable to the needs of *Homo sapiens* – was taken away? This is the question we address in Chapter 9.

9

A Climate of Peril

In 2005, as Hurricane Katrina crashed into New Orleans, Andre decided to stay at home with his wife and young child.[1] In the six years since they had moved in, their sturdy brick house had not flooded once, despite the many storms that had hit the city. But as the levees broke and New Orleans was inundated with roughly 250 billion gallons of water, Andre and his family were forced up to the second storey of their home. Even there, the water level continued to rise, stopping only when it reached waist height. Walking out onto his balcony, Andre was greeted by the sound of neighbours trapped inside attics yelling out for help. The next morning, he and his family were transported to an overpass where they would wait without food or water for a further two nights, before eventually paddling an air mattress to his sister's house uptown. While Andre and his family lost their home, they escaped with their lives. Many inhabitants of New Orleans – 1,833 in total – were not so lucky.

In 2017, Cape Town faced precisely the opposite challenge, coming dangerously close to running out of water entirely as a result of a severe and sustained drought.[2] In 2019, as wildfires tore through Australia over the Christmas holiday, Sydney was blanketed in a smoky haze, with air pollution reaching more than ten times hazardous levels.[3] That same year, wildfires in California briefly put San Francisco among the most polluted cities on the planet.[4] In 2022, the heatwave that swept across Europe saw temperatures in London surpass 40 °C, the highest ever recorded in the city.[5] Rail tracks buckled under the heat, and a runway at Luton Airport melted. Most of the city stayed at home in a desperate bid to keep cool and hydrated. Over just three days, roughly a thousand people across Britain died from complications related to the heat.[6] The same year also saw Pakistan experience record temperatures followed by record rains, leading to one-third of the country being flooded, killing thousands and displacing 33 million people, about 15 per cent of the population.[7]

These and countless other examples attest to the reality that climate change is already upon us. Denial has become an untenable position in light of both hard data and lived experience. Disasters like Hurricane Katrina are no longer an abnormality: warming ocean temperatures are making storm winds more intense by increasing air moisture levels, while rising sea levels are exacerbating storm surges.[8] Nine of the ten hottest years on record all occurred in the past decade, causing heatwaves, droughts and wildfires.[9] And unless action on decarbonization is rapidly accelerated, recent events will pale in comparison to the devastation that lies ahead.

The world's cities are poorly positioned for the effects of climate change. 90 per cent of urban areas today are coastal, placing many cities at risk of being engulfed, especially as storms become more severe.[10] Around 800 million people live in urban coastal areas where sea level rise is projected to reach at least 0.5 metres by 2050.[11] Paved surfaces not only ease the movement of flood waters, but also magnify heat, raising city temperatures above the surrounding countryside. The fact that cities are often also located in basin areas makes them traps for the debris wafted in the air by wildfires. Constant flooding, unbearable heat and toxic air are what the future has in store for many cities.

This chapter will make clear the magnitude of the threat that climate change poses to our urban world, and the actions that must urgently be taken. Across many countries, efforts are already underway to adapt cities for a more hostile climate. While these are a step in the right direction, much more will be needed. And in the rapidly growing cities of the developing world, a lack of resources is limiting the available options. More importantly, adaptation will simply not be enough to indefinitely prevent a collapse in standards of living. Decarbonization is our duty to future generations. Cities already account for most of the world's emissions, and their share will rise further as the global urbanization rate continues to climb. Cities must therefore be front and centre in efforts to avert the catastrophe we are hurtling towards.

A HISTORY OF VULNERABILITY

Humankind's endeavour to tame its environment has long been part of the story of the city. Recall from Chapter 2 how the management of irrigation systems and flood defences was one of the original reasons that cities came into existence. Without such protections from the capriciousness of rainfall, the development of human civilization would have been constantly set back.

Yet through most of history *Homo sapiens* remained largely at the mercy of the planet. The Indus Valley civilization, home to the ancient cities of Mohenjo-Daro and Harappa, is the first recorded example of a complex society collapsing under the pressures of climate change. Prior to around 3000 BCE, the ferocity of the summer monsoon in the Indus Valley area made it unsuitable for agricultural settlement.[12] That began to steadily change as a result of a process known as the axial precession, a 26,000-year cycle in the earth's orbital rotation that impacts the relative severity of seasonal contrasts in the northern and southern hemispheres. Farming in the region soon became possible as floods declined in severity, and by around 2500 BCE a thriving civilization had emerged. As the monsoon continued to weaken, however, rivers dried up and the land became increasingly arid, unable to support the agricultural surplus necessary to feed its cities.[13] By 1800 BCE the cities of the Indus Valley had been abandoned and civilization had unwound.

The collapse of the Maya tells a similar story. By 400 BCE, Mayan society had developed into an interconnected network of city-states across Central America. Between 800 and 1000 CE, however, that civilization collapsed during a period of intense and recurrent droughts.[14] The cause of this intense dry spell was a southward shift in the system of winds that historically brought rainfall over Central America during the middle of the year, a shift in which the earth's axial precession may have again been a factor.[15] Deteriorating growing conditions meant the region was no longer able to support the Maya, whose population had expanded significantly over the preceding centuries. Conflict between city-states, probably fuelled by competition for scarce resources, is thought to have accelerated the decline.[16]

Perhaps the most dramatic example of the toll of climate change in human history was the simultaneous collapse of multiple Bronze Age civilizations in the Eastern Mediterranean starting around 1200 BCE, which we mentioned briefly in Chapter 2. The New Kingdom of Egypt, the Hittite Empire of Anatolia, and the early Minoan and Mycenaean civilizations of Greece all collapsed in the course of little over a century. A sharp increase in northern hemisphere temperatures ushered in a period of widespread drought that brought not only famine to the Eastern Mediterranean, but also invading armies from beyond the region.[17] These mysterious 'Sea People', who were themselves possibly fleeing deteriorating environmental conditions, laid waste to the already-weakened civilizations of the region.[18]

Since industrialization, humanity has become far more resilient to climatic shifts. Modern engineering means rivers can be redirected and water drawn up from the earth in a way that never before was possible. Technological innovations have dramatically increased total food production, while the globalization of food supply means shortfalls in one region can, if trade works and importers can afford it, be offset by surplus production in another. Access to food remains tragically unequal around the world, but sustained shortages of the type that once triggered civilizational collapse have not occurred for a long time.

Yet the progress of the past two centuries has also set us along a path towards a climate potentially far more volatile than anything seen in human history. Fossil fuels have been at the heart of this, powering humankind's transition from a world of rural subsistence to one of urban abundance. That has made us far richer, but is now undermining the environmental foundations on which human progress has relied for millennia.

THE DANGERS AHEAD

Cities are a testament to humanity's mastery over the natural world. With concrete, steel and bitumen, we bend and shape the physical environment to our will. But these monuments to human ingenuity are far more vulnerable than many realize.

Consider Miami. Close to 25 million visitors make the trip to this wonderfully vibrant beachfront city in a typical year. But Miami faces a challenging future as a result of climate change. Much of the city's waterfront is on track to be submerged by rising sea levels. Figure 3 below illustrates how the city is on course to ultimately end up submerged as a result of global warming. King tides that flood the streets of downtown Miami are already becoming more frequent.[19] Increased tropical storm severity is set to make the city a progressively more dangerous place to be in during the hurricane season.

Shanghai offers another example of the threat that rising sea levels poses to cities. The city's position at the meeting point of the Yangtze River and the East China Sea has long made it a critical hub for trade in and out of China, but its low elevation now places it in jeopardy from flooding. Much of the city's downtown, along with both of its airports and a number of major landmarks, are at risk of being swallowed up by the encroaching waters.[20]

Many cities exposed to flooding as a result of climate change are already taking measures to protect themselves. Miami is building sea walls and installing pumps to transfer water back out to sea. Shanghai has installed a vast drainage system and is in the process of constructing over 120 miles of flood-prevention

FIGURE 3: Rising sea levels will leave many cities under water.
Long-run sea level rise under different temperature-rise scenarios

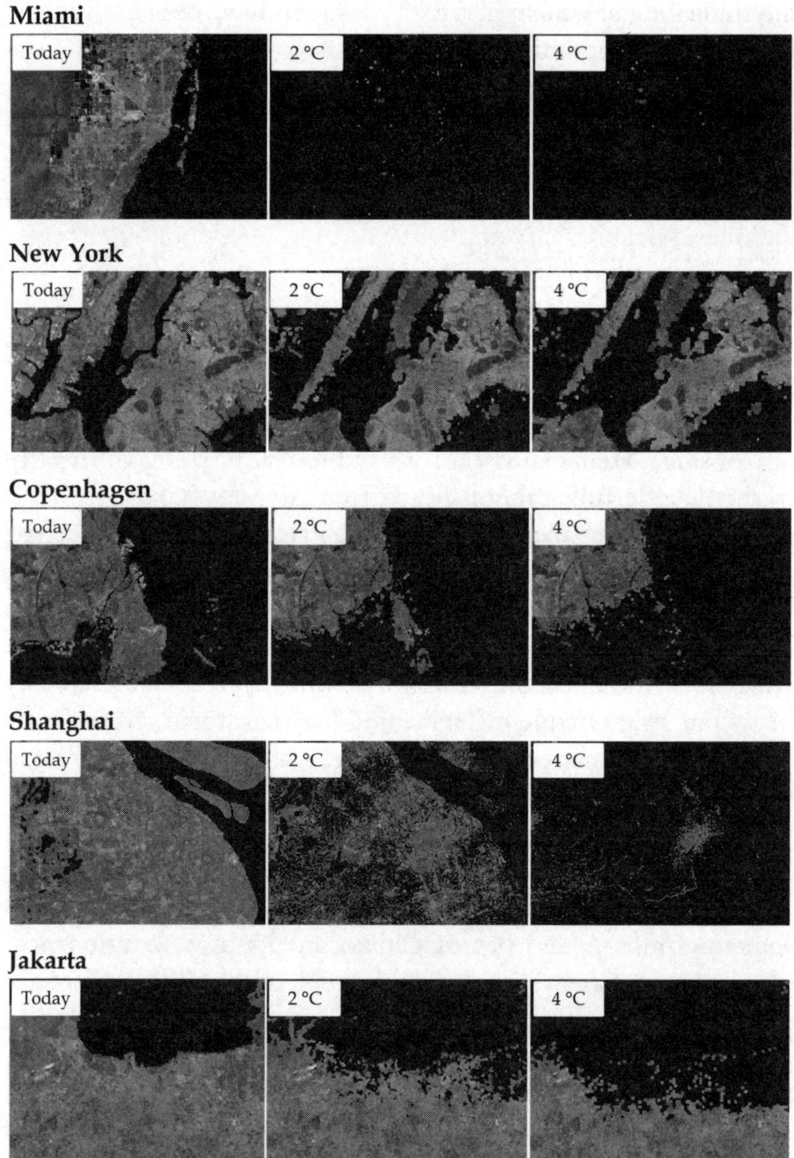

Source: EarthTime

walls.[21] Rotterdam, another city at risk of submersion, has spent decades preparing itself. 20 years ago, it completed construction on a gigantic storm-surge barrier known as the Maeslantkering that shields it from the North Sea.[22] Since then, the city has been constructing parks, plazas and other spaces that can serve as temporary reservoirs that absorb flooding.[23]

Around the world, a number of low-lying coastal cities are also sinking, compounding the threat from rising sea levels. Jakarta, Indonesia's capital city and home to over 10 million inhabitants, is one such example. Not only is the city located right on the Java Sea, but it is also built on swampy land interspersed by a network of rivers. The extraction of groundwater by the local population combined with rapid construction is causing the city to sink by 25 centimetres a year in the northern area closest to the waterfront.[24] By 2050, it is estimated that a third of the city could be under water.[25] And Jakarta is not an isolated case. Bangkok, Manila, Ho Chi Minh City and many other poor cities are rapidly sinking. As noted in Chapter 2, floodplains, with their more fertile soil, have historically been a magnet for the large human settlements that coalesce into cities. But as fields are replaced by concrete, such geography turns from an asset to a liability.

In the Netherlands, where low elevation has long demanded protection against encroaching waters, defensive infrastructure has been built over hundreds of years. The pace of climate change today simply does not allow for this. And while cities like Miami may with sufficient national support have the resources to invest in fast-tracking major engineering projects to reduce their exposure, most cities in poor countries do not. Low-cost strategies do exist, such as elevating houses on stilts or reintroducing coastal mangrove swamps to act as a natural barrier against storm surges. And for sinking cities, providing a reliable municipal water supply can reduce the pace of groundwater extraction. But these actions alone are unlikely to protect the inhabitants of such cities from being displaced by worsening floods. The fact that the government of Indonesia is planning to relocate its capital city from Jakarta to the island of Borneo highlights the gravity of the threat.

Heat stress is another major challenge on the horizon for cities. In London, heatwaves have historically been greeted with enthusiasm. Revellers flock to the city's parks, perhaps stopping off for an ice cream along the way, and joyfully embrace a respite from the city's typically dreary weather. The youthful and confident strip down to their bathers to work on their tans. In recent years, as temperature peaks have become more extreme, that has changed.

Cities are exposed to what is known as the urban heat island effect. While trees and vegetation deflect radiation from the sun, roads and buildings do the opposite, absorbing and retaining heat. The issue is compounded by the so-called waste heat that is generated by cars and machinery. During the summer, the temperature in Paris and other major cities can be as much as 10 °C higher than in the surrounding countryside.[26]

In the cities of the rich world, many urban dwellers are able to insulate themselves from oppressive heat through air-conditioning, even if few homes in historically cooler cities such as London are equipped with that technology today. Indeed, air-conditioning has made it possible for rich countries to build cities like Phoenix or Dubai in the middle of deserts, albeit at enormous environmental cost. In hot and humid Singapore, the technology is almost universal, and considered essential to the city's productivity. When asked what he thought the most important technological innovation of the twentieth century was, the city's founding father Lee Kuan Yew picked the air-conditioner.[27]

In the cities of the developing world, air-conditioning is a luxury few can afford. Beyond the loss of life, inhabitants of cities such as Delhi or Lagos are likely to suffer economically as a result of working hours lost to heat stress. Many street vendors and construction workers will be forced to seek shelter during the hottest hours of the day, leaving them in an even more precarious financial position than what they experience today.

Increased heat will also worsen the already lethal air pollution in many developing world cities. Globally, air pollution is responsible for several times more deaths than war, terrorism and murder combined, and most of these deaths occur in poor countries.[28] Wildfires like those described earlier in the chapter are only one way in which climate change will exacerbate the problem. Heatwaves make possible the chemical reactions that turn various air pollutants into ground-level ozone, a substance that is especially toxic for humans. Delhi has long had a problem with hazardous air quality, which is responsible for over 50,000 deaths ever year in the city.[29] Heatwaves in recent years have increased the number of days in which ozone has exceeded safe levels, making the problem even more severe.[30]

In the daily news we are already getting a glimpse of the suffering that lies ahead for people living in the cities of the developing world. Consider Sufia, who lives in the slum of Duaripara in Dhaka with her husband and three children.[31] Sufia's makeshift tin house is unbearable in the summer heat. The ceiling fan they have installed does little to help, merely circulating the same hot air, and their access to electricity is patchy anyway. The low-lying area in which Sufia's slum is located is regularly flooded. When that happens, Sufia and her family are forced to sit on top of a bed and wait for the water to recede. For people like Sufia, climate change threatens their already-precarious existence.

A PERFECT STORM

With over 4,500 square miles of glacial ice, the Himalayan mountain range is sometimes referred to as earth's Third Pole.[32] The Indus River, the Ganges, the Mekong, the Yangtze and the Yellow River are all fed by these glaciers. Altogether, nearly two billion people live in the basins of rivers that originate in the Himalayas.[33]

Rising temperatures are causing the glaciers of the Himalayas to recede rapidly. If temperatures rise above 2 °C, at least two-thirds of these glaciers will disappear.[34] The catastrophic effect will play out in two phases. First, as the ice melts, the rivers that originate in the mountains will experience recurrent flooding, imperilling lives and livelihoods across much of Asia. Once that has happened, water availability will decline precipitously as these glaciers cease to feed melting ice into the rivers. An increased rural exodus into already-stretched cities like Dhaka is almost certain to follow, as subsistence farmers face starvation in the countryside. Increased frequency and severity of heatwaves will compound rural hardship, as high temperatures and extreme weather kill crops and livestock and make it impossible for farmers to work.

Meanwhile, in Africa, the semi-arid Sahel region that borders the Sahara Desert is at risk of becoming uninhabitable. Droughts are becoming more frequent and more severe, while the rainy season is becoming both shorter and more torrential.[35] As the dry earth is unable to absorb the periodic heavy rains, flooding washes away crops and livestock.[36] Already oppressive heat is getting worse. In the decades ahead, deteriorating conditions will force tens of millions of people southwards towards cities like Nairobi and Lagos. Those cities, which are already struggling to cope with the challenges of urban poverty described in Chapter 7, will themselves need to contend with recurrent flooding, oppressive heat and toxic air quality as a result of climate change.

As climate change increases the pressure on people in vulnerable areas, tensions will mount. Countries of the Sahel, such as Mali and Niger, have already become hotspots for Islamic extremists who feed off discontent. Increasing resource scarcity will also raise the prospect of conflict between nations. Access to the waters of the Indus River and its tributaries has been a historical source of strain between India and Pakistan, and a major reason that both sides claim the disputed Kashmir region in the north.[37] As water becomes scarcer, the strained relationship between the two nations could deteriorate further.

Through a combination of environmental pressures and the associated erosion of livelihoods and increased conflicts over land, water and other scarce resources, climate change will displace countless numbers of people in poor countries. The overwhelming majority will end up in the already-straining cities of the developing world, in their own and neighbouring poor countries. Some portion, however, will make their way to North America, Europe or other rich regions in pursuit of a better and safer life for themselves and their families. The refusal of many rich countries to accept more refugees will come under growing pressure, as it becomes even clearer that remaining at home is not an option for millions more people. If recent debates over Syrian refugees in Europe are anything to go by, the issue is likely to become divisive and vitriolic. But rich nations cannot in good conscience seal themselves off from the fate of their fellow human beings, not least as their fossil-fuelled growth is a major contributor to the problem. Instead, there must be a level-headed debate over how more refugees and migrants can be accepted, and how they should be integrated.

URBAN SOLUTIONS

The risks posed by climate change demand urgent and dramatic action to slow the emission of carbon and other greenhouse gases. Cities represent just over 55 per cent of the world's population today, but directly and indirectly account for 70 per cent of emissions.[38] By 2050, that share could be 85 per cent, given projections for further increases in the global urbanization rate. Finding ways to reduce the carbon footprint of cities will be pivotal in the fight against climate change. Switching the energy grid from coal and gas to solar, wind and other renewable sources is an essential part of that, but it is not the whole answer.

Cities generate emissions in two ways. The first is direct emissions, which come from the activities that occur within a city. That includes driving cars, heating and cooling buildings, and keeping the power on. It includes, further, all the emissions generated by factories operating within city boundaries. The inhabitants of cities also consume many products that originate elsewhere, particularly food harvested in rural areas, as well as cement, concrete, steel, cars and other carbon-intensive products. These are what are known as the indirect emissions, and make up an important additional component of the carbon footprint of cities.

The reason why cities account for a disproportionate share of global emissions is that they tend to be much richer than rural areas. With a higher standard of living typically comes more emissions. When adjusting for income, cities are actually *less* emissions-intensive than rural areas, accounting for 80 per cent of global GDP but only 70 per cent of total emissions. The decisive factor in this is density, and in particular the impact it has on the emissions generated by transportation. In Britain, inhabitants of rural areas generate one and a half times the emissions per capita from transport.[39] In the US, where rural communities are often sparsely populated, it is 2.3 times.[40] That is why escaping to the country is not necessarily beneficial for the environment.

Equally, there is significant scope for cities to make themselves less emissions-intensive and, at the same time, more liveable. The first is to reduce sprawl. Living in the suburbs (or exurbs) tends to involve relying heavily on your car, not only to get to work, but also to get to shops, restaurants and most other places. That results in a higher carbon footprint compared with inner cities.[41,42] This is one reason why the highly suburbanized cities of the US 'sunbelt' – places like Atlanta or Houston – produce a lot more emissions per capita than cities like New York or Boston, an issue compounded by underdeveloped public transit systems in those places.[43] (The other factor behind higher emissions in sunbelt cities is their heavy reliance on air-conditioning, which is used to cope with high temperatures.)

Sprawl is both bad for the environment and bad for quality of life. As described in Chapter 4, many people living at the periphery of cities today would rather be located more centrally, but simply cannot afford it, particularly once they start having children. A large part of the answer to this lies in increasing the housing supply in urban cores, through, for example, more mid-rise development and the repurposing of disused industrial and office spaces. That will make proximity to the urban centre more affordable. As discussed in Chapter 5, the trend towards remote work also creates an opportunity to bring the urban back to suburban areas, as demand for amenities like cafes and gyms moves closer to home. Reimagining suburbs as walkable neighbourhoods that support hybrid workers during the days they are not in the office would help further reduce carbon footprints.

In addition to reducing the distances residents need to travel, cities can also lower their carbon footprint by encouraging cleaner modes of transport. A medium-sized car with a single passenger produces just over 300 grams of carbon-equivalent emissions per mile, whereas a typical train journey produces a little over 65 grams per passenger mile.[44] Cycling and walking produce effectively zero emissions. Access to alternative modes of transport varies widely across cities. In New York, more than half of workers commute by public transit, while a further one in ten either bike or walk to work.[45] In Houston, on the other hand, fewer than one in twenty workers rely on public transit for their commutes, and almost nobody cycles or walks.[46] Increasing the coverage and reducing the cost of public transport, as well as creating more options for cyclists and pedestrians, needs to be a priority. Beyond reducing emissions, switching to these alternative modes of transport will also improve air quality, free up urban parking space for redevelopment, and make cities more accessible for those who cannot afford their own car.

Beyond shifting the mix of transport, cities also need to embrace electrification using clean sources. Oslo, the capital city of Norway, offers an instructive example. In recent years, the city has made significant strides in electrifying its public transport system while also encouraging inhabitants to switch their own petrol cars for electric vehicles (EVs). In fact, the city now claims the highest number of EVs per capita in the world.[47] Government subsidies combined with public investment in charging infrastructure have been key to that. And the fact that 98 per cent of electricity in Norway comes from renewable sources means that EVs do not simply shift emissions to the power grid.[48] Elsewhere, in cities like London, Paris and Barcelona, the transition to EVs is being accelerated by the introduction of fees for driving petrol or diesel vehicles through certain zones. That is a positive first step, but it needs to be combined with greater reliance on renewable energy sources. Another example of what is possible is Ljubljana, the capital of Slovenia, where investments by the government in public transport and cycle routes have made a car-free city a reality.[49]

Making buildings more energy-efficient is another major area of opportunity for cities, with the potential to reduce emissions and save money at the same time. Solar panels can be affixed to urban rooftops. Households can be equipped with smart meters to better understand their energy usage. Homes, offices, shops and the like can be insulated to reduce temperature variability. District heating systems can capture waste heat produced at industrial sites and use it to warm buildings throughout the city.

Adding more green space to cities is also important. The late biologist Edward Osborne Wilson, who we first introduced in Chapter 2, coined the term 'biophilia' to describe our hardwired attraction to other forms of life.[50] Being in nature has a deeply calming effect on us.[51] That is perhaps one reason why so many people still voluntarily choose to live in rural areas despite the challenges they can bring. But cities need not be places devoid of nature. Integrating green spaces into cities not only makes them nicer places to live, but also helps to reduce their environmental footprint. Trees provide a natural mechanism for capturing carbon emitted into the air. And by counteracting the urban heat island effect, green spaces help to reduce reliance on artificial methods of cooling.

The presence of greenery varies significantly from city to city. Seattle has 885 parks. Nashville, by contrast, has just 243.[52] As a result, nearly every resident in Seattle lives within ten minutes' walking distance from a park, while less than half do in Nashville.[53] Even US cities like Seattle lag behind the global leaders in this area. More than 40 per cent of the land of Singapore is covered with greenery, the result of a decades-long commitment that began with the city's founder Lee Kuan Yew.[54] The city has over 30 square miles of green space – an area larger than Manhattan – linked together by over 200 miles of tree-covered walking and cycling paths.[55] In recent years, Singapore's 'Gardens by the Bay' development has emerged as a major tourist attraction in its own right, as visitors flock to see the towering artificial trees that act as both vertical gardens and solar power generators. Singapore is a high-density city, yet it manages to continue freeing up space for green developments like this. Paris has already set out plans to follow suit, with a 2030 target for 50 per cent of the surface area of the city to be covered by planted areas, including everything from parkland to rooftop gardens.[56]

The actions we have discussed so far all focus on the direct emissions produced by cities, but there is also much cities can do to lower their indirect emissions as well. Reducing consumption by wasting less is a good start. New York City and its surrounding suburbs dispose of over 30 million tonnes of waste every year, or roughly 1.5 tonnes per person, most of which ends up in landfill.[57] As waste breaks down, it emits methane, which contributes directly to emissions. But the bigger problem is all the greenhouse gases that were emitted in generating those products in the first place. For example, agriculture is a major contributor to greenhouse gas emissions, but roughly one-quarter of all the world's food is wasted.[58] While part of that comes from spoilage in the supply chain, households, retailers and restaurants also play a big role, particularly in rich countries.[59] Beyond disposing of less waste in the first place, more can be done to increase recycling rates. This is not just about households more diligently separating out their trash. For instance, the carbon footprint of the construction industry could be significantly reduced by recycling more old concrete, which can be crushed and used for paving or as aggregate for mixing into fresh concrete mixes. Altogether, about 75 per cent of waste generated in the US could be recycled, but only about 30 per cent actually is.[60,61] That means there is roughly 100 million tonnes of waste going into landfill in the US every year that could have been recycled, and the vast majority of that is generated in cities. Recycling all that waste would be the equivalent of taking 60 million cars off the road.[62] At a minimum, cities should start charging households and businesses based on the amount of rubbish they contribute to landfill.

Another intriguing possibility for reducing the indirect carbon footprint of cities is moving agriculture from rural to urban areas. With vertical farming, a space the size of a supermarket can replace about 280 hectares of farmland, while artificial lighting and temperature control can greatly increase yields.[63] Establishing vertical farms within the periphery of urban areas would free up a significant amount of agricultural land for carbon-absorbing forests, while also reducing emissions from transportation and food spoilage. Today the prospects for vertical farming are hampered by high energy requirements. If powered by renewables, however, a lettuce grown in a vertical farm could end up generating 70 per cent fewer emissions than one grown in an open field.[64]

While many cities in rich countries have made reducing emissions a priority, most in poor countries simply lack the resources to do so. The problem is that, even if rich countries were to stop all greenhouse gas emissions tomorrow, the world would still be on a path towards catastrophic climate change. Together, rich countries account for around 60 per cent of the stock of carbon emissions that have accumulated in the atmosphere since the onset of the industrial revolution, but only around 30 per cent of the annual flow of additional emissions each year, and their share of each annual additional contribution is rapidly falling.[65] If countries like India or China reach the level of emissions per capita of the US, we will have no chance of stopping climate change. Poor countries need to follow a less emissions-intensive path to development, and rich countries must support them financially to achieve that. Generating power using more renewable energy is one part of the solution, but designing the fast-growing cities of the developing world in a way that limits demand for fossil fuels will also be key.

Better global coordination is also required to ensure that developing countries do not become dumping grounds for emissions-intensive economic activity. China is now the largest emitter of carbon in the world, but a meaningful portion of that comes from the manufacturing of products ultimately consumed in places like London or New York.[66] As rich countries have offshored manufacturing over recent decades, they have also offshored a substantial share of their carbon emissions, which makes progress on emissions reduction look better than it has been in practice.[67] Climate change is a fundamentally global problem, and it demands a global perspective if action is to be successful.

OUR FUTURE IN THE BALANCE

The longer we delay action on climate change, the closer we get to a point of no return at which a series of interconnected climate processes will become self-reinforcing and lead to a spiralling crisis. These fall into three types.[68] The first relates to the planet's ice sheets, and involves a self-reinforcing cycle in which melting ice exposes darker surfaces of ground or ocean. These surfaces absorb more heat, which in turn accelerates ice melting. In many areas, these ice sheets have covered vegetation that has steadily broken down over thousands of years, creating large amounts of methane ready for release into the atmosphere. The second mechanism relates to biospheres such as forests and oceans, and occurs when the trees or algae that absorb carbon can no longer sustain the critical mass they need to survive in a warming environment. The third relates to the patterns of ocean and air circulation that dictate much of the earth's climate, but could potentially be disrupted by warming temperatures and melting ice.

In the century ahead, we may see the East Antarctic ice sheet melt, the Amazon rainforest turn to grassland, and the Gulf Stream collapse, to name just a few of the many possibilities.[69] Each of these would be catastrophic. Together, they could be apocalyptic. Poor countries will be hit hardest at first, but rich countries will not be able to insulate themselves indefinitely. It is not our wish to be alarmist, but it is vital to recognize that the risks presented by climate change are truly disastrous.

The global population may have become more resilient to climatic changes, but as we look ahead to the end of this century and beyond, the shifts in our natural environment that will result if climate change proceeds unchecked would be unlike anything seen since the dawn of human civilization. In Chapter 2, we explored how the transition to the benign environment of the Holocene made possible a virtuous circle of human progress that began with the first cities. There is no guarantee that progress will survive in the world we are creating for future generations.

Many readers may already feel desensitized by the continuous onslaught of predictions about climate change. The time for more information passed long ago; now is the moment for action. We are in a race against climate change, and right now we are dawdling at the start line.

10

Conclusion: Better Together

In 1800, there were 1 billion humans sharing our planet, roughly 70 million of whom inhabited cities. Today, the global population is 8 billion, with over 4.5 billion people living in cities. Not only are we more numerous and clustered together, but we are also bound by a fast-flowing stream of goods, people and information that wraps around the entire planet. Lewis Mumford once wondered whether the world would eventually be transformed into one 'vast urban hive'.[1] We are not far from it. At no point in history has there been so dramatic a change in so short a time in the condition of *Homo sapiens*.

Around the middle of the twentieth century, global population growth began to slow down. By the end of the twenty-first century, as birth rates continue to fall, the great boom in human population brought about by declining mortality rates will come to an end, with the United Nations forecasting the global population to plateau at a little over 10 billion people.[2] Even as population growth tapers off and eventually reverses, the share of the world's population living in cities is expected to continue rising, driven by the ongoing urbanization of the developing world. The result is that, by the end of the century, the total population living in cities may have doubled from today's levels.[3]

We therefore are still a long way away from the end of urban growth. And we have many questions still to answer. How will cities accommodate all these extra people? Where will they find employment? How will we protect ourselves from climate change, pandemics and other risks? It is naïve to imagine we will stumble upon the answers to these questions by relying blindly on market forces and individual choice. To make our way successfully through this transition, we need to shape our destiny, not simply take our chances.

We should not fool ourselves into believing there are easy solutions. Catchy slogans and wishful thinking will merely hold back progress. The urban world is a complex system, prone to unintended consequences. Incremental solutions and continuous experimentation are required, grounded in patience and a long-term perspective. Policy choices cannot be made in isolation from one another, and need to be considered in their wider context.

In what follows, we pull together the many lessons of this book into an agenda for the decades ahead to make our urban world fairer, more cohesive and more sustainable. This agenda is everybody's business. All levels of government – from municipalities to multilateral organizations – must play their part in delivering it. Businesses and other non-governmental institutions like universities need to be brought into the fold. And all our readers, every individual, has a crucial role to play, by living more sustainably, participating more in community life, and choosing leaders who take these matters seriously.

OVERHAUL URBAN DESIGN

Since at least the time of Hippodamus, the founder of urban planning in Greece, humans have been experimenting with different ways to design cities to make them functional and enjoyable places to live and work. (Hippodamus's own contribution was superimposing a street grid onto the city to ensure that any destination could be reached from any starting point without a lengthy detour.)

It's time to admit that the great twentieth-century experiment in car-based sprawl was not only a colossal waste of resources, but also a social and environmental failure. Cars are a wonderful convenience for the individual, but a terrible basis for mass transportation. They not only pollute the air and heat the atmosphere, but also take up an immense amount of space. In the US, parking absorbs fully one-third of all the space in cities, with an estimated eight parking spots for every car.[4] Moving people around by car rather than shared transport modes like bus or rail is also incredibly inefficient, given that private cars spend 95 per cent of their time parked.[5]

It could have been so different. The more people use public transit, the more accessible and convenient it becomes, as services can be run with greater frequency along a wider range of routes. Just imagine the kind of public transit systems we could have built if all the money we spent on cars and roads in the twentieth century had been pooled into shared modes of transport.

The irony is that the convenience provided by cars is self-eroding. As people become reliant on driving for a growing share of journeys, the amount of traffic congestion increases. New roads help to ease that congestion for a time, but quickly lead to what urban planners call induced demand, as faster journey times encourage people to drive even more. It takes just a few years for the typical metropolitan highway to fill up with increased demand.[6]

What cars have brought is privacy and choice. They have made it possible for the majority of urban populations in rich countries to move out to the suburbs, where they can live in large homes in quiet cul-de-sacs with their own backyards. And they have made it possible for people to shuttle from destination to destination without ever having to encounter anyone not of their choosing. But seclusion and insularity are poor foundations for a cohesive society.

The challenges associated with low-density sprawl have been understood for a long time. In her pioneering 1961 book

The Death and Life of Great American Cities, Jane Jacobs argued that the subdivision of cities into areas with distinct functions – for living, for working and for playing – made them less safe and less enriching places to live.[7] Since then, many cities have embraced mixed-use development and encouraged the regeneration of inner city neighbourhoods. Urban centres from Copenhagen to Vancouver have reclaimed streets for pedestrians, helping to increase the casual encounters that foster a sense of community. London has added bike lanes while Sydney has reintroduced streetcars, helping inhabitants in central urban neighbourhoods to shake their dependence on driving and reduce their carbon footprint. Paris has embraced the principle of a '15-minute city', which suggests that inhabitants should be able to find everything they need – including work, food, healthcare, education and culture – within a 15-minute walk or bike ride.

Such initiatives are a step in the right direction, but they leave many challenges unresolved. As high-paid knowledge workers have returned to inner cities in recent decades, they have enjoyed the fruits of these efforts. Meanwhile, low-paid service workers have been relegated to distant commuter zones where much of their free time is absorbed either stuck in traffic or navigating whatever patchwork of public transport options they have available to them. Cities need to think bigger, reimagining not just the urban core, but the entire sprawling network that the modern city has morphed into over recent decades.

Improving housing affordability will be a critical part of that. Cities need to prioritize adding housing supply in central urban areas where prices are least affordable, not at the periphery of cities where it is already cheap. Otherwise, inner cities will eventually become the exclusive playgrounds of the rich. Even New York City still has plenty of scope for adding mid-rise development. Manhattan may well feel like a sea of skyscrapers, but roughly 70 per cent of buildings across the five boroughs are three storeys high or less.[8] The repurposing of disused office buildings, industrial spaces and parking lots also offers a significant opportunity to increase housing supply in inner cities. Further densification would have the added benefit of freeing up space for more parkland and urban farms, making cities not only more pleasant but also more environmentally friendly.

As cities go about densification, they should also expand their stock of social housing to help create more socioeconomically integrated communities. With the rapid increase in rental prices over recent decades, there are solid grounds for raising the threshold at which social housing can be accessed. The experience of Vienna shows that having a broader segment of the income distribution living in these developments would also help to mitigate the negative effects of concentrated poverty.

In addition to making the urban core more accessible to those on lower incomes, cities also need to extend the principles of mixed-use development and walkability to the suburbs. The idea of the city as a series of concentric circles with distinct functions does not have to be taken as an article of faith. Anyone who doubts that suburbs can be transformed need only look at the history of London's Chelsea or Paris's Montmartre, both of which were once suburban. To kickstart the process, cities could identify areas like abandoned malls or languishing high streets and offer tax breaks to hospitality and retail businesses to encourage them to set up shop. Constructing terraced houses and even mid-rise developments near to these precincts could help to provide a critical mass of customers within easy walking or cycling distance.

Bringing public transport systems into the twenty-first century is central to this transformation. Networks needs to be electrified, based on renewable sources, and extended into suburban and exurban areas. That can be paid for at least in part by introducing or increasing congestion charges for cars. That would have the added benefit of nudging people to rely more on public transport, which in turn would improve its viability. Reliance on public transport could also be increased by shifting its revenue base away from ticket sales and towards general taxation, making it cheaper for users. Pricing models like London's that place a heavy penalty on those living further from the city also need to be reviewed, given the centrifugal movement of poverty away from inner city areas.

REBUILD FOR THE KNOWLEDGE ECONOMY

It is also time to accept that there has been a permanent and irreversible shift away from manufacturing jobs in rich countries. Even if a meaningful share of manufacturing activity is repatriated from low-cost locations to rich countries, high levels of automation mean the sector will not reclaim its historical role as a ladder of upward economic mobility for those without a higher education. The same can be said for the clerical work that was offshored in recent decades and is now increasingly being digitized and replaced by artificial intelligence.

We have been too slow to act on these trends. The power of specialization makes global trade a potent mechanism for increasing the world's total productive capacity, making goods and services cheaper and more abundant. And technological change is the bedrock of rising living standards over the long run. But both also create significant friction as workers are forced to adjust to shifting patterns in the availability of jobs. Initiatives like the Trade Adjustment Assistance Program in the US, which focuses on providing workers displaced by globalization with financial support for retraining and relocation, have simply been far too modest to have any meaningful impact.

As traditional middle-class jobs have disappeared, many once prosperous cities and towns in the United States, Britain, France and elsewhere have entered into spirals of decline, trapping their inhabitants in generations of poverty. The fact that many living in these places have turned to populist leaders is understandable: the system no longer works for them.

Ignoring the places left behind by globalization and automation is no longer tenable. Looking ahead, a new agenda for a more inclusive economy is required, one that blends vision with pragmatism, acknowledging that no government can defy the laws of gravity that pull people together in the knowledge economy.

The experience of cities like Seattle or Leipzig shows there is hope for reversing the self-perpetuating cycles of joblessness and decay that plague many formerly prosperous cities. To rebuild struggling cities for the knowledge economy, governments need to come at the problem from two directions. First, they need to increase high-skill employment opportunities in the area, either by making it easier to start a new business or by offering incentives such as tax breaks for existing businesses to establish local operations. The latter should be treated with caution, given the potential for a race to the bottom between cities that ends up benefiting only shareholders. A better strategy is to selectively target a few businesses to serve as anchors for the local economy. Second, cities need to bring in the skilled workers that employers are seeking. Investing in universities is a good start, but is often not enough to keep workers around. Creating the type of urban environment that knowledge workers increasingly want to live in – dense and lively, not sprawling and lifeless – is also important. As the city builds a foundation of high-paid knowledge jobs, opportunities in the non-tradeable services that support them will also grow and local government finances will improve.

To sustain a less centralized economy, long-distance transportation infrastructure needs to be modernized in many rich countries. The invention of the railway was critical in supporting the dispersion of economic activity during industrialization. Often those very same rail networks are still the fastest way to get between many cities. Countries like

Britain need to commit to high-speed rail, but they need to do it in the right way. Simply connecting Manchester to London risks drawing an even greater share of knowledge workers to the capital, particularly if hybrid working only requires them to be in Manchester for part of the week. Ensuring that Manchester is an attractive place for those workers to spend their leisure time would partially help to reduce that risk. But it is also important to consider how the multiple cities of the North could be knit together to create a metropolitan system that can serve as a counterweight to London.

The forces of agglomeration are strong, which is why it is important to maintain a pragmatic perspective on 'levelling up' struggling places. The economies of Britain and France need not implode entirely into their capital cities, but it is also not realistic to expect a long list of small cities and towns to thrive in a global economy in which scale is increasingly important for competitiveness.

Relocation is hard. Friendships fall apart, treasured places are lost, and family is far away. It is also expensive, which is why providing mobility vouchers to help those whose jobs have been displaced is a smart idea. The bigger problem, however, is that the economics often simply do not stack up, as the higher cost of living in many thriving cities – particularly when it comes to housing – more than offsets the potential for higher wages for less skilled workers. That leads many to stay put in left-behind places, which only compounds the problem, as too many people chase too few low-skill jobs. This is why making cities like London, New York and San Francisco more affordable is so important.

And whether you live in a rich or poor neighbourhood in a rich or poor place, all children must be given access to a good education. A failure to do so is not only profoundly unfair, but also deeply inefficient, as so much latent potential is wasted. Access to education will never be levelled in countries like the United States unless the funding of schools is decoupled from local economic conditions. With education one of the best investments a society can make in today's knowledge economy, it's high time for a new deal on how it is funded.

Given the pace of technological change today, there is a need to rethink the full lifecycle of education. The idea of a stage of learning followed by a stage of working followed by a stage of retirement is now obsolete. Rather than retrenching workers when they no longer have the requisite skills, companies should be incentivized to retain, reskill and, if necessary, relocate them for new roles. Shifting responsibility for reskilling to employers will increase their openness to cheaper methods of learning, such as micro-credentials, that today are often not accepted as a valid alternative to a degree, in turn lowering the overall cost burden to society. Universities should play a proactive role in partnering with companies to assist them in the delivery of such programmes. Embedding the reskilling process within firms will also reduce the likelihood that displaced workers embark on the process of gaining a new skillset only to find themselves unable to secure a job at the end of it.

ACCELERATE SUSTAINABLE DEVELOPMENT

Nearly all of the growth in the global urban population over the remainder of this century will occur in developing economies. Financial and technical support from rich countries is needed to help these cities develop into thriving centres of opportunity for their inhabitants, rather than catchments of desperation as bulging populations flee deteriorating conditions in the countryside. That support is particularly important given the added challenge of developing along a less emissions-intensive path than the one followed by the cities of the rich world.

The continued growth of informal settlements lacking in the most basic services is not only a blight on our collective conscience, but a significant risk to humanity. Overcrowded and unsanitary conditions in these settlements will make them catalysts for the global spread of new infectious diseases. Investing to improve health systems and upgrade informal settlements is part of the solution, but it is also critical to address the underlying causes of the deep poverty in many cities in the developing world.

As automation makes it increasingly challenging for poor countries to follow the traditional path of industrial development, a significant expansion of investment in education will be required. While we are sceptical that the future of knowledge work is entirely remote, we do see scope for certain activities to be shifted to well-educated workers in poor countries in cases where the gains to companies from lower wages overseas outweigh the disadvantages of remote collaboration. Data science, financial analysis and doubtless other areas of knowledge work could probably be performed from abroad. The experience of Bangalore suggests that a hub for offshoring knowledge work can also mature into a global centre of innovation in its own right.

It is vital that, as they become more prosperous, the cities of the developing world do not adopt the car-based sprawl of many rich world cities. The world simply cannot afford more cities with the carbon footprint of sprawling Houston. Prioritizing public transport over private cars in the development of urban infrastructure is critical, as is incorporating adequate green space throughout these cities.

Cities like Bogota are already showing an impressive degree of leadership in this area. When Enrique Peñalosa was elected mayor of the Colombian capital in 1997, he set about reorienting the city away from cars by cutting back on planned highway expansions, introducing a public bus system, rolling out cycle paths and prioritizing mixed-use development.[9] Not only was this about lowering the city's carbon footprint and reducing its notorious levels of air pollution; it was also about transforming the city to be more equitable for its many residents who simply could not afford a car.

Many poor countries will not be able to establish prosperous and sustainable cities without significant support from rich countries, and certainly not at the speed that is required. Only a handful of donor countries – Denmark, Norway, Sweden, Germany and Luxembourg – currently give at or above the target set by the United Nations of 0.7 per cent of national income.[10] For a time Britain also delivered to that target, but in 2021 it took the regressive step of slashing its foreign aid budget. Now is not the moment for rich countries to turn inwards, given the increasingly global nature of the challenges confronting us. The $100 billion of annual support pledged in 2015 at the Paris Conference to help developing countries reduce their emissions and adapt to climate change is a good start, but with each passing year the gap between the commitments and delivery in this and other areas of development assistance is becoming more galling.

DELIVER LOCALLY, SUPPORT NATIONALLY, COORDINATE GLOBALLY

There is no escaping the fact that this agenda will require an active role for government. That will bring with it challenges of its own, including the frustrations of political gridlock and the prospect of taxpayer resources being wasted on ill-conceived projects or redirected based on political rather than economic criteria.

One way to mitigate these challenges is to adopt the principle of subsidiarity, which suggests that responsibility – and accountability – should not be shifted to more centralized levels of government unless there is a clear rationale for doing so. Less centralized levels of government generally tend to have a better understanding of a community's needs and fewer conflicting interests to manage. The role of centralized government is to redistribute resources across communities and manage issues that it would be duplicative or impractical to handle at a more local level.

City governments cannot manage issues like national defence, but they can and should have a major role in delivering the agenda we have described in this book. This is likely to lead to better decisions and faster progress. For that to work, however, city government jurisdictions need to be redefined to encompass both urban core and periphery. The expansion of the modern city has meant that many metropolitan areas are now comprised of multiple local governments each responsible for one geographic area, often without any kind of mechanism to align on a combined urban plan and allocate investments where they are most needed. The San Francisco Bay Area has muddled through by forming an association for the multiple local governments that operate in the area. Britain has taken the sensible step of creating an overarching mayoralty for its larger cities such as London and Manchester, though their power and resources remain limited. Many cities still need to find a workable solution to this challenge.

For this to work, city governments need to expand their breadth and depth of capabilities. During the Covid-19 pandemic, cities varied widely in their ability to detect outbreaks, implement and enforce social distancing measures, and roll out vaccines. Those that struggled should be investing now to ensure they are in a better position to handle future pandemics, and they need help from national and international authorities to do so. In areas from health to education to transport and beyond, competent municipal governments that are appropriately resourced and attract capable and motivated leaders are essential.

There is much that city governments cannot do on their own. Those that have fallen into cycles of economic decline need help from national and, where applicable, state governments to muster the investment they require to transform themselves. While in the short run this may disadvantage the inhabitants of well-off cities who end up donating their tax contributions to struggling places, helping such places rise to their potential will benefit everyone in the long run through a more productive national economy. Supporting others to thrive is not a zero-sum game.

Equally, it is important to ensure that redistributed resources are used wisely. As mentioned earlier, trade-offs are going to be required in a knowledge economy in which the scale of a city matters for competitiveness. Unfortunately, national funds are often allocated across local districts in a seemingly arbitrary fashion based on who has the most political leverage at any given point in time. That is why defining a multi-year national economic development plan with a clear and transparent set of investment priorities is so important. This is the approach Japan has followed for decades, and it has been a crucial part of that country's ability to maintain low levels of income disparity between cities.

The global nature of challenges such as climate change and pandemics means that coordination across borders is essential. In recent years, growing geopolitical hostilities have complicated this. While it may not be a substitute for bilateral and multilateral diplomacy, there is a valuable supporting role in this environment for inter-city diplomacy. One example of this is the C40 Cities partnership on climate change. The group brings together the leaders of nearly 100 of the world's largest cities across almost 50 countries – including both China and the United States – around a shared pact to reduce their emissions in line with the ambition to limit global warming to 1.5 °C. In the decades ahead, we expect to see cities increasingly asserting themselves as major players on the international stage. City networks have a long and illustrious history. The Hanseatic League, the confederation of Northern European cities introduced in Chapter 2, not only established commercial ties over vast distances, but also helped defend its member cities from banditry and common enemies. Such networks lost their potency as national governments came to dominate the political landscape. While nation-states are here to stay, city networks can and must play a supporting role in solving global challenges given the fraying of international relations that is currently underway. More than 300 such inter-city networks already exist today.[11]

At the extreme, some argue that cities should be set free from the constraints of national governments. One such example of this is the charter city movement, originally championed by Nobel Prize-winning economist Paul Romer. The idea of the charter city movement is to grant municipal governments the right to develop their own system of governance and undertake policy reforms free from the constraints of the existing political system. Enacted in full, it would constitute a significant step back towards the fragmented political landscape of city-states that once prevailed. It is an unrealistic idea. Independent city-states like Singapore and Dubai have thrived as connecting points in global trade networks. Others, like Luxembourg, have benefited by operating as tax havens. That is not a scalable model for the global economy. It is also not the way to resolve the grave challenges currently confronting the world. What we need is multiple layers of government – local, national and multilateral – acting in concert with one another.

In a 1987 interview with *Woman's Own* magazine, British prime minister Margaret Thatcher once declared: 'There is no such thing as society', only 'individual men and women and families'.[12] We beg to differ. *Homo sapiens* is a social creature, and our collective prosperity depends on the strength of the bonds between us. Since its first emergence five millennia ago, the city has been the ultimate expression of these bonds.

A dangerous set of challenges confronts our world today, with cities at their centre. For the more than four billion people who currently live in them – and the billions more who will join them in the coming decades – it is vital that cities thrive. Cities are drawing together the collective capabilities of humankind at ever-greater scale. But that is no cause for complacency. The speed with which change is occurring has accelerated sharply. The speed with which we continually adjust our course must increase too.

We all have a role to play in this. Change starts with our own behaviour, everything from what we eat to how we travel to who we spend time with. Stepping up and taking initiative matters too, from the local all the way to the national level.

Cities, with their unbounded creative potential, provide a source of hope for the future. By working together to improve them, we can create a better life for all.

ACKNOWLEDGEMENTS

We could not have wished for a more encouraging and helpful editor than Tomasz Hoskins at Bloomsbury Continuum. From his early enthusiastic embrace to detailed editing, his guidance and support has been invaluable. Thanks also to the project editor at Bloomsbury, Sarah Jones; to Nick Fawcett, who copyedited the manuscript; to the proofreader, Guy Holland; and to Rachel Nicholson and Sarah Head in publicity and marketing respectively.

We have been fortunate to receive feedback from a number of leading experts in the topics covered in this book. For generously taking time from their busy schedules to offer us their wisdom, we would like to thank: Ed Glaeser, Chairman of the economics department at Harvard and one of the world's leading experts on the economics of cities; Peter Frankopan, Professor of Global History at Oxford and a renowned authority on the Silk Roads; John-Paul Ghobrial, Professor of Global and Modern History at Oxford and a valued colleague of Ian's at Balliol College; Lord Norman Foster, not only a world-renowned architect but also an insightful observer of the built environment; Billy Cobbett, director of Slum Dwellers International with a lifetime of experience working to understand and improve cities in developing countries; Robert Muggah, co-author with Ian of *Terra Incognita: 100 Maps to Survive the Next 100 Years* and an expert on all things cities; and Carl Frey, Director of the Oxford Martin Programme on the Future of Work and an innovative economic historian. This book has been vastly improved by their contributions.

We have also benefited from the work of a number of excellent research assistants. Harry Camilleri provided tremendously rich and wide-ranging research assistance for an earlier iteration of this book and helped us improve the manuscript. Oliver Purnell developed an immensely useful survey of the issues and data that provided a foundation for our later work. Rachel Barber facilitated the recruitment of research assistants.

Oxford Economics graciously provided complimentary access to their excellent database on global cities, which helped inform much of the original analysis in this book. We also benefited greatly from the brilliant prior work done on many of the topics covered in this book by the Our World in Data team at the Oxford Martin School.

None of these individuals or institutions are accountable for the arguments in this book, or for any errors or omissions, which we alone take responsibility for.

Ian Goldin and
Tom Lee-Devlin

In writing this book I have benefited from residencies at the magnificently situated and perfectly managed Rockefeller Center in Bellagio and the Stellenbosch Institute for Advanced Study. At both I was provided with the time to think and write undisturbed and benefited from conversations with fellow residents across many disciplines.

I could not wish for a better work environment than that provided by the Oxford Martin School at the University of Oxford. My research programmes on The Future of Work and Technological and Economic Change are funded by Citi, and my programme on The Future of Development by the Allan & Gill Gray Foundation. Their generosity has enabled me to employ an outstanding group of post-doctoral fellows, together with a programme manager and assistant. While not directly involved in this book, their support has allowed me to manage a highly active research programme while writing. My college, Balliol, continues to nourish me intellectually and provide a ready source of expertise for my curious mind.

My family have once again provided the emotional foundation and supportive cocoon that I have needed. I am immensely grateful to Tess, who with extraordinary patience and forbearance has tolerated the endless nights and weekends I have devoted to working on what is my 24th book. Her loving support is what makes my writing possible.

Tom Lee-Devlin's insights, writing skills and ability to distil nuggets from mountains of information are evident on every page of this book, and I am delighted to have had him as a co-author.

Ian Goldin, Oxford

Many hours writing this book were spent in the wonderful surrounds of the London Library, which provided both a serene place to work and an excellent research collection.

My wife Megan has been a constant encouragement throughout the long process of writing. I am also thankful for Michael Goldman, Patrick Keenan and other friends who provided valued pointers on the topics we have covered. And I am grateful for the inspiring colleagues I am surrounded by at *The Economist*, from whom I learn so much about the world and how to explain it.

Finally, it has been a privilege to work with Ian Goldin on this book, who has been both an inspiring thought partner and a generous and patient mentor.

Tom Lee-Devlin, London

NOTES

Chapter 1 – Introduction

1 Ritchie, H. and Roser, M., 2019, 'Urbanization', Our World in Data (ourworldindata.org).
2 Ibid.
3 Ibid.
4 Ibid.
5 Goldin, I. and Muggah, R., 2020, *Terra Incognita: 100 Maps to Survive the Next 100 Years* (Century), p. 97.
6 Mumford, L., 1961, *The City in History* (Harcourt).
7 Ibid.
8 Ibid.
9 Augustine, 426, *The City of God: Volume I* (reprint by Jazzybee Verlag, 2012, accessed online at Perlego.com).
10 Rousseau, J.-J., 1762, *Emile, or On Education: Book I* (reprint by Heritage Books, 2019, accessed online at Perlego.com).
11 Jefferson, T., 1800, *From Thomas Jefferson to Benjamin Rush*, National Archives (archives.gov).
12 Hazlitt, W., 1837, *Characteristics in the manner of Rochefoucault's Maxims* (J. Templeman).
13 Howard, E., 1898, *Garden Cities of Tomorrow* (reprint by Perlego, 2014, accessed online at Perlego.com).
14 Authors' calculations based on US Census data.
15 Authors' calculations based on Oxford Economics data.
16 Florida, R., et al., 2008, 'The rise of the mega-region', *Cambridge Journal of Regions, Economy and Society*, Vol. 1, No. 3.
17 Authors' calculations based on Oxford Economics data.
18 Mumford, *The City in History*.
19 Zhao, M., et al., 2022, 'A global dataset of annual urban extents (1992–2020) from harmonized nighttime lights', *Earth System Science Data*, Vol. 14, No. 2.

20 Ritchie, H. and Roser, M., 2019, 'Land use', Our World in Data (ourworldindata.org).
21 Schwedel, A., et al., 2021, 'The working future: more human, not less', Bain & Company (bain.com).
22 Roser, M., Ritchie, H. and Ortiz-Ospina, E., 2015, 'Internet', Our World in Data (ourworldindata.org).
23 Goos, M. and Manning, A., 2007, 'Lousy and lovely jobs: the rising polarization of work in Britain', *The Review of Economics and Statistics*, Vol. 89, No. 1.
24 Spengler, O., 1922, *The Decline of the West, Volume Two* (reprint by Arktos Media Ltd, 2021, accessed online at Perlego.com).
25 Abel, J. and Deitz, R., 2019, 'Why are some places so much more unequal than others?', *Economic Policy Review*, Vol. 25, No. 1.
26 Lindsey, R. and Dahlman, L., 2022, 'Climate change: global temperature', National Oceanic and Atmospheric Administration (climate.gov).
27 NASA (Vital Signs of the Planet), 'Key indicators: Arctic sea ice minimum extent' (climate.nasa.gov).
28 World Meteorological Organization, 2021, *Atlas of Mortality and Economic Losses from Weather, Climate and Water Extremes (1970–2019)* (wmo.int).
29 *The Economist*, 2021, 'The pandemic's true death toll'.
30 For more detail see Goldin and Muggah, *Terra Incognita*, pp. 106–10.

Chapter 2 – Engines of Progress

1 Goldin and Muggah, *Terra Incognita*.
2 Norwich, J., 2009, *The Great Cities in History* (Thames & Hudson).
3 Wilson, B., 2020, *Metropolis* (Jonathan Cape).
4 Ibid.
5 Grimbly, S., 2013, *Encyclopedia of the Ancient World* (Fitzroy Dearborn).
6 Ibid.
7 Wood, M., 2020, *The Story of China* (Simon & Schuster).
8 Grimbly, *Encyclopedia of the Ancient World*.
9 Bairoch, P., 1988, *Cities and Economic Development* (University of Chicago Press), p. 6.
10 Ibid.

11 Lieberman, B. and Gordon, E., 2021 (2nd edition), *Climate Change in Human History* (Bloomsbury).
12 Bairoch, *Cities and Economic Development.*
13 Wilson, E. O., 2012, *The Social Conquest of the Earth* (Liveright).
14 Dunbar, R., 1992, 'Neocortex size as a constraint on group size in primates', *Journal of Human Evolution*, Vol. 22, No. 6.
15 Wood, *The Story of China.*
16 Wilson, *Metropolis.*
17 Ibid.
18 Harari, Y. N., 2014, *Sapiens* (Harper).
19 Mumford, *The City in History.*
20 Bellah, R., 2011, *Religion in Human Evolution* (Harvard University Press).
21 George, A., 1999, *The Epic of Gilgamesh* (Penguin Classics), p. 4.
22 Hall, P., 1998, *Cities in Civilization* (Weidenfeld & Nicolson).
23 Ibid.
24 Van De Mieroop, M., 2021, *A History of Ancient Egypt* (Wiley).
25 Turchin, P., 2009, 'A theory for formation of large empires', *Journal of Global History*, Vol. 4, No. 2.
26 Ibid.
27 Whitehouse, H., et al., 2022, 'Testing the Big Gods hypothesis with global historical data: a review and "retake"', *Religion, Brain & Behavior.*
28 Ibid.
29 Bekhrad, J., 2017, 'The obscure religion that shaped the West', BBC (bbc.com).
30 Haverfield, F., 1904, *The Romanization of Roman Britain* (reprint by Perlego, 2004, accessed online at Perlego.com).
31 Kuhn, D., 2009, *The Age of Confucian Rule* (Harvard University Press).
32 Fessenden, M., 2015, 'Making a sandwich from scratch took this man six months', *Smithsonian Magazine.*
33 Gwendolyn, L., 2001, *Mesopotamia: The Invention of the City* (Penguin).
34 Mumford, *The City in History*, p. 104.
35 Squitieri, A. and Altaweel, M., 2018, *Revolutionizing a World: From Small States to Universalism in the Pre-Islamic Near East* (UCL Press).

36 Berube, A. and Parilla, J., 2012, 'Metro trade: cities return to their roots in the global economy', Brookings Institution (brookings.edu).
37 Frankopan, P., 2015, *The Silk Roads* (Bloomsbury).
38 Verbruggen, R., et al., 2014, 'The networked city', in Knox, P. (ed.), *Atlas of Cities* (Princeton University Press).
39 Frankopan, P., 2015, *The Silk Roads* (Bloomsbury).
40 Barjamovic, G., et al., 2019, 'Trade, merchants, and the lost cities of the Bronze Age', *The Quarterly Journal of Economics*, Vol. 134, No. 3.
41 Berube and Parilla, 'Metro trade'.
42 Sanghani, N., 2019, 'Lessons from the history of globalisation', Capital Economics (capitaleconomics.com).
43 Roser, M., Ritchie, H. and Ortiz-Ospina, E., 2019, 'World population growth', Our World in Data (ourworldindata.org).
44 Mumford, *The City in History*.
45 Johnson, S., 2010, *Where Good Ideas Come From* (Penguin).
46 Al-Khalili, J., 2012, *The House of Wisdom* (Penguin).
47 Sachs, J., 2020, *Ages of Globalization* (Columbia University Press).
48 Dunbar, K., 1997, 'How scientists think: on-line creativity and conceptual change in science', in Ward, T. B., et al. (eds), *Conceptual Structures and Processes: Emergence, Discovery, and Change* (American Psychological Association Press).
49 Ferguson, N., 2017, *The Square and the Tower* (Penguin).
50 Ibid.
51 Mokyr, J., 2016, *Culture of Growth: The Origins of the Modern Economy* (Princeton University Press).
52 Chandler, T. and Fox, G., 1974, *3000 Years of Urban Growth* (Academic Press, accessed online at Perlego.com).
53 Clossick, J., 2014, 'The industrial city', in Knox, P. (ed.), *Atlas of Cities* (Princeton University Press), p. 75.
54 Chandler and Fox, *3000 Years of Urban Growth*.
55 Lieberman and Gordon, *Climate Change in Human History*.

Chapter 3 – Levelling Up

1 Wood, J., 1893, *Dictionary of Quotations: From Ancient and Modern, English and Foreign Sources* (Frederick Warne and Co., accessed online at guntenberg.org).

2 Chandler and Fox, *3000 Years of Urban Growth*.
3 Kim, S. and Margo, R., 2003, 'Historical perspectives on US economic geography', NBER working paper.
4 Paullin, C. and Wright, J., 1932, *Atlas of the Historical Geography of the United States* (reprint by Greenwood Press, 1976).
5 Kim and Margo, 'Historical perspectives on US economic geography'.
6 Boustan, L., et al., 2013, 'Urbanization in the United States, 1800–2000', NBER working paper.
7 Ibid.
8 Meyer, D., 1989, 'Midwestern industrialization and the American manufacturing belt in the nineteenth century', *The Journal of Economic History*, Vol. 49, No. 4.
9 Authors' calculations based on historical US Census Bureau data.
10 Gibson, C., 1998, *Population of the 100 largest cities and other urban places in the United States: 1790 to 1990* (US Census Bureau).
11 Purdy, H., 1945, *A historical analysis of the economic growth of St Louis, 1840–1945* (Federal Reserve Bank of St Louis).
12 Longman, P., 2015, 'Why the economic fates of America's cities diverged', *The Atlantic*.
13 Authors' calculations based on historical US Census Bureau data.
14 Ibid.
15 Boustan, et al., 'Urbanization in the United States, 1800–2000'.
16 Longman, 'Why the economic fates of America's cities diverged'.
17 Authors' calculations based on Oxford Economics data.
18 Boustan, et al., 'Urbanization in the United States, 1800–2000'.
19 Schwedel, et al., 'The working future'.
20 Goldstein, A., 2017, *Janesville: An American Story* (Simon & Schuster).
21 Federal Reserve Bank of St Louis, 2019, 'The US auto labor market since NAFTA' (stlouisfed.org).
22 Acemoglu, D. and Restrepo, P., 2019, 'Automation and new tasks: how technology displaces and reinstates labor', *Journal of Economic Perspectives*, Vol. 33, No. 2.
23 Schwedel, et al., 'The working future'.
24 Marshall, A., 1890, *Principles of Economics* (reprint by Nabu Press, 2010).

25 Jaffe, A. B., et al., 1993, 'Geographic localization of knowledge spillovers as evidenced by patent citations', *The Quarterly Journal of Economics*, Vol. 108, No. 3.
26 Berkes, E. and Gaetani, R., 2021, 'The geography of unconventional innovation', *The Economic Journal*, Vol. 131, No. 636.
27 Schwedel, A., et al., 2020, 'Peak profits', Bain & Company (bain.com).
28 Goldin, I., et al., 2022, 'Why is productivity slowing down', *Journal of Economic Literature* (forthcoming).
29 Longman, 'Why the economic fates of America's cities diverged'.
30 Ibid.
31 Authors' calculations based on US Census data and Oxford Economics data.
32 Ayers, D., 2019, 'The gender gap in marriages between college-educated partners', Institute for Family Studies (ifstudies.org).
33 Moretti, E., 2010, 'Local multipliers', *American Economic Review*, Vol. 100, No. 2.
34 Ibid.
35 Sturge, G., 2022, 'Migration statistics', House of Commons Library (commonslibrary.parliament.uk).
36 Kerr, S. and Kerr, W., 2018, 'Immigrant entrepreneurship in America: evidence from the survey of business owners 2007 & 2012', NBER working paper.
37 Ibid.
38 Authors' calculations based on US Census data.
39 Frost, R., 2020, 'Are Americans stuck in place? Declining residential mobility in the US', Joint Centre for Housing Studies of Harvard University (jchs.harvard.edu).
40 Diamond, R. and Moretti, E., 2021, 'Where is standard of living the highest? Local prices and the geography of consumption', NBER working paper.
41 Authors' calculations based on Nationwide Bank data.
42 Authors' calculations based on National Centre of Education Statistics data.
43 Federal Reserve Bank of New York, 2021, 'The labor market for recent college graduates' (newyorkfed.org).
44 Moretti, E., 2012, *The New Geography of Jobs* (First Mariner Books), p. 76.

45 Jain, V., 2019, 'Case study on territorial development in Japan', World Bank (worldbank.org).
46 Miwa, N., 2022, 'High-speed rail development and regional inequalities: evidence from Japan', *Transportation Research Record.*
47 Authors' calculations based on Oxford Economics data.
48 Puga, D., 2002, 'European regional policies in light of recent location theories', *Journal of Economic Geography*, Vol. 2, No. 4.
49 Elliott, L., 2019, 'HS2 would widen UK north–south divide and should be axed, says report', *Guardian.*
50 Moretti, *The New Geography of Jobs.*
51 Glaeser, E., 2005, 'Reinventing Boston: 1630–2003', *Journal of Economic Geography*, Vol. 5, No. 2.
52 Shapiro, S., 2015, 'New species of city discovered: university city', Next City.
53 Florida, R., 2002, *The Rise of the Creative Class* (Basic Books).
54 Matsu, J., et al., 2022, 'Investing in regional equality – lessons from four cities', Chartered Institute of Public Finance and Accountancy (cipfa.org).
55 Ibid.
56 Ibid.
57 Ibid.

Chapter 4 – Divided Cities

1 Jowett, B., 1888, *The Republic of Plato: Translated into English* (Oxford University Press, accessed online at Perlego.com).
2 Autor, D., 2019, 'Work of the past, work of the future', *AEA Papers and Proceedings*, Vol. 109.
3 Song, X., et al., 2019, 'Long-term decline in intergenerational mobility in the United States since the 1850s', *Proceedings of the National Academy of Sciences,* Vol. 107, No. 1.
4 Ibid.
5 de Tocqueville, A., 1958 (written 1833–5), *Journeys to England and Ireland* (reprint by Routledge, 2017, accessed online at Perlego .com).
6 Clossick, 'The Industrial City', in Knox (ed.), *Atlas of Cities*, p. 78.
7 Engels, F., 1845, *The Condition of the Working Class in England* (reprint by Perlego, 2005, accessed online at Perlego.com).

8 Clossick, 'The Industrial City', in Knox (ed.), *Atlas of Cities*, p. 78.
9 Sinclair, U., 1906, *The Jungle* (reprint by Open Road Media, 2015, accessed online at Perlego.com).
10 Clossick, 'The Industrial City', in Knox (ed.), *Atlas of Cities*.
11 Florida, R., 2017, *The New Urban Crisis* (Basic Civitas Books).
12 *The Dearborn Independent*, 1922, *Ford Ideals* (Dearborn Publishing Company).
13 Huber, M., 2013, *Lifeblood: Oil, Freedom, and the Forces of Capital* (University of Minnesota Press), p. 73.
14 Morphet, J., 2013, 'A city-state?', in Bell, S. and Paskins, J. (eds), *Imagining the Future City: London 2062* (Ubiquity Press).
15 de Vise, P., 1976, 'The suburbanization of jobs and minority employment', *Economic Geography*, Vol. 52, No. 4.
16 Hobbs, F. and Stoops, N., 2002, 'Demographic trends in the 20th century', US Census Bureau (census.gov).
17 Cited in Wilkerson, I., 2016, 'The long-lasting legacy of the Great Migration', *Smithsonian Magazine*.
18 Wilson, W., 1987, *The Truly Disadvantaged* (University of Chicago Press).
19 Rajan, R., 2019, *The Third Pillar* (William Collins).
20 Hannah-Jones, N., 2019, 'It was never about busing', *New York Times*.
21 Clark, W., 'School desegregation and white flight: a re-examination and case study', *Social Science Research*, Vol. 16, No. 3.
22 Rhodes, J. and Brown, L., 2019, 'The rise and fall of the "inner city": race, space, and urban policy in postwar England', *Journal of Ethnic and Migration Studies*, Vol. 45, No. 17.
23 Ehrenhalt, A., 2013, *The Great Inversion* (Vintage).
24 Edlund, L., et al., 'Gentrification and the rising returns to skill', NBER working paper.
25 Authors' calculations based on Zillow data and US Census population weightings.
26 Autor, D., et al., 2017, 'Gentrification and the amenity value of crime reductions', NBER working paper.
27 Gropius, W., 1969, 'Principles of Bauhaus production', in Wingler, H. (ed.), *The Bauhaus* (MIT Press), pp. 109–10.
28 Brooks, D., 2000, *Bobos in Paradise* (Simon & Schuster).
29 Glass, R., 1964, *London: Aspects of Change* (MacGibbon & Kee), p. xix.

30 Easterly, W., et al., 2016, 'A long history of a short block: four centuries of development surprises on a single stretch of a New York city street' (greenestreet.nyc/paper).
31 Authors' calculations based on ONS data.
32 Ibid.
33 Clark, D., 2022, 'Average age of mothers at childbirth in the United Kingdom from 1938 to 2020', Statista (statista.com).
34 Dougherty, C. and Burton, A., 2017, 'A 2:15 alarm, 2 trains and a bus get her to work by 7 a.m.', *New York Times*.
35 Jeunesse, E., 2017, 'The rise of poverty in suburban and outlying areas', Joint Center for Housing Studies (jchs.harvard.edu).
36 White, G., 2015, 'Long commutes are awful, especially for the poor', *The Atlantic*.
37 Samans, R., et al., 2015, 'Inclusive growth and development report 2015', World Economic Forum (weforum.org).
38 McFarland, J., et al., 2018, 'The condition of education 2018', National Centre for Education Statistics (nces.ed.gov).
39 Semuels, A., 2017, 'Japan might be what equality in education looks like', *The Atlantic*.
40 Ibid.
41 Fryer, R., 2017, 'Management and student achievement: evidence from a randomized field experiment', NBER working paper.
42 Weale, S., 2021, 'Headteachers call for reform of school admissions to redress attainment gap', *Guardian*.
43 Shelter, 2022, 'The story of social housing' (england.shelter.org.uk).
44 Jacobs, J., 1961, *The Death and Life of Great American Cities* (reprint by Vintage, 1993).
45 *The Economist*, 2020, 'Governments are rethinking the provision of public housing'.
46 Shelter, 2022, 'Social housing deficit' (england.shelter.org.uk).
47 Ball, J., 2019, 'Housing as a basic human right', *The New Statesman*.
48 Authors' calculations based on ONS data.
49 Bonnet, F., 2014, 'Social housing in New York', Metropolitics (metropolitics.org).
50 *The Economist*, 2020, 'Governments are rethinking the provision of public housing'.
51 Kober, E., 2020, 'New York City records another year of jobs–housing mismatch', Manhattan Institute (economics21.org).

52 The *Guardian* Data Blog, 2016, 'Deprivation and poverty in London', *Guardian*.
53 Stoll, B., et al., 2022, 'Benchmarking European cities on creating the right conditions for zero-emission mobility', Clean Cities Campaign (cleancitiescampaign.org).
54 Transport for London, 2022, 'Fares from 1 March 2022' (tfl.gov.uk).
55 RATP, 2022, 'Tickets and fares' (ratp.fr).

Chapter 5 – Remote Work: The Threat to Cities

1 Nilles, J., 1976, *The Telecommunications–Transportation Tradeoff: Options for Tomorrow* (John Wiley & Sons).
2 Toffler, A., 1980, *The Third Wave* (Bantam Books).
3 Walters-Lynch, W., 2020, 'What 50 years of remote work teaches us', BBC (bbc.com).
4 *The Economist*, 2021, 'The rise of working from home'.
5 Ibid.
6 Broom, D., 2021, 'Home or office? Survey shows opinions about work after Covid-19', World Economic Forum (weforum.org).
7 Schwedel, et al., 'The Working Future'.
8 Bloom, N., Twitter, 29 August 2022, 14:59.
9 Westfall, C., 2021, 'Mental health and remote work: survey reveals 80% of workers would quit their jobs for this', Forbes (forbes.com).
10 Nuffield Health, 2020, 'Working from home taking its toll on the mental health & relationships of the nation' (nuffieldhealth.com).
11 Schwedel, et al., 'The Working Future'.
12 Gibbs, M., et al., 2021, 'Work from home & productivity: evidence from personnel & analytics data on IT professionals', *SSRN Electronic Journal*.
13 Cross, R., et al., 2021, 'Collaboration overload is sinking productivity', *Harvard Business Review*.
14 Allen, T., 1977, *Managing the Flow of Technology* (MIT Press).
15 Waber, B., et al., 2014, 'Workspaces that move people', *Harvard Business Review*.
16 Bloom, N., et al., 2015, 'Does working from home work? Evidence from a Chinese experiment', *The Quarterly Journal of Economics*, Vol. 130, No. 1.
17 D'Onfro, J., 2015, 'Steve Jobs had a crazy idea for Pixar's office to force people to talk more', *Business Insider*.

18 Bloom, et al., 'Does working from home work?'
19 Morgan, K., 2021, 'Why in-person workers may be more likely to get promoted', BBC (bbc.com).
20 University College London, 2017, 'Audience members' hearts beat together at the theatre' (ucl.ac.uk).
21 Winck, B., 2021, '4 charts show how fast everyone is flocking back to big cities', *Business Insider*.
22 Anderson, D., 2021, 'Home prices in cities rise 16%, surpassing suburban and rural price growth for the first time since pre-pandemic', Redfin (redfin.com).
23 Transport for London, 2022, 'Latest TfL figures show continued growth in ridership following lifting of working from home restrictions' (tfl.gov.uk).
24 *The Economist*, 2022, 'The future of public transport in Britain'.
25 Kastle Systems, 2022, 'Recreation activity vs. office occupancy' (kastle.com).
26 Bloom, Twitter.
27 Partridge, J., 2022, 'Google in $1bn deal to buy Central Saint Giles offices in London', *Guardian*.
28 Putzier, K., 2021, 'Google to buy New York City office building for $2.1 billion', *Wall Street Journal*.
29 Morris, S. and Hammond, G., 2022, 'Citi plans £100m revamp of Canary Wharf tower', *Financial Times*.
30 Hartman, A., 2021, 'Salesforce says "the 9-to-5 workday is dead" and will provide 3 new ways for employees to work – including the possibility of working from home forever', *Business Insider*.
31 Gupta, A., et al., 2022, 'Work from home and the office real estate apocalypse', NBER working paper.
32 Cohen, B., 2022, 'Yes, Zoom has an office. No, it's not a place to work', *Wall Street Journal*.
33 Cadman, E., 2021, 'Your new office set-up is going to look a lot like the old one', Bloomberg News.
34 Ibid.
35 Chernick, H., et al., 2011, 'Revenue diversification in large U.S. cities', IMFG Papers on Municipal Finance and Governance.
36 Hammond, G., 2021, 'Canary Wharf: does the east London office district have a future?', *Financial Times*.
37 Hammond, G., 2022, 'Canary Wharf proposes £500mn lab project to reinvent financial hub', *Financial Times*.

38 Cintu, A., 2021, 'Record apartment conversions make 2021 most successful year in adaptive reuse', RentCafe (rentcafe.com).
39 Feldman, E., 2022, 'Converting office buildings to apartments could help city housing crisis', *Spectrum News*.
40 Ehrenhalt, *The Great Inversion*.
41 Reagor, C., 2017, 'Downtown Phoenix's rebirth has been decades in the making. Here's how they did it', *The Arizona Republic* (azcentral.com).
42 Kelly, H., Kramer, A. and Warren, A., 2020, *Emerging Trends in Real Estate: United States and Canada 2020*, PWC and Urban Land Institute.
43 Harmon, J. and Zim, E., 2021, 'The rise and fall of the American mall', *Business Insider*.
44 Authors' calculations based on Zillow data.
45 Paulas, R., 2017, 'The death of the suburban office park and the rise of the suburban poor', *Pacific Standard Magazine*.
46 DeVoe, J., 2021, 'Leaders say Arlington has the ecosystem to be the next Austin or Miami', ARL Now.
47 Authors' calculations based on Zillow data.

Chapter 6 – Cities, Cyberspace and the Future of Community

1 Turner, F., 2006, *From Counterculture to Cyberculture* (University of Chicago Press), p. 13.
2 Authors' calculations based on General Social Survey data.
3 Rajan, *The Third Pillar*.
4 Putnam, R., 2000, *Bowling Alone* (Simon & Schuster).
5 Ibid.
6 Haidt, J., 2022, 'Why the past 10 years of American life have been uniquely stupid', *The Atlantic*.
7 McLuhan, M., 1962, *The Gutenberg Galaxy* (University of Toronto Press).
8 Marvin, C., 1988, *When Old Technologies Were New* (Oxford University Press), pp. 199–200.
9 Ibid.
10 Deibert, R., 1997, *Parchment, Printing, and Hypermedia: Communication in World Order Transformation* (Columbia University Press).

11 Anderson, B., 1983, *Imagined Communities* (Verso).
12 Prior, M., 2007, *Post-Broadcast Democracy: How Media Choice Increases Inequality in Political Involvement and Polarizes Elections* (Cambridge University Press), p. 1.
13 Pool, I., 1983, *Forecasting the Telephone* (Praeger).
14 Fischer, C., 2007, 'Technology and society: historical complexities', *Sociological Inquiry*, Vol. 67, No. 1.
15 Chen, W., 2013, 'Internet use, online communication, and ties in Americans' networks', *Social Science Computer Review*, Vol. 31, No. 4.
16 Haidt, 'Why the past 10 years of American life have been uniquely stupid'.
17 Van Alstyne, M. and Brynjolfsson, E., 1997, 'Electronic communities: global village or cyberbalkans', *Economic Theory*.
18 Sunstein, C., 2007, 'Neither Hayek nor Habermas', *Public Choice*, Vol. 134.
19 Tempest, K., 2021, *On Connection* (Faber & Faber), p. 61.
20 Bor, A. and Bang Petersen, M., 2021, 'The psychology of online political hostility: a comprehensive, cross-national test of the mismatch hypothesis', *American Political Science Review*, Vol. 116, No. 1.
21 Ibid.
22 Prapotnik, K., et al., 2019, 'Social media as information channel', *Austrian Journal of Political Science*.
23 Grant, A., Twitter, 29 August 2021, 15:03.
24 Cho, J., et al., 2020, 'Do search algorithms endanger democracy? An experimental investigation of algorithm effects on political polarization', *Journal of Broadcasting & Electronic Media*, Vol. 64, No. 2.
25 Nordbrandt, M., 2021, 'Affective polarization in the digital age: testing the direction of the relationship between social media and users' feelings for out-group parties', *New Media & Society*.
26 Rathje, S., et al., 2020, 'Out-group animosity drives engagement on social media', *Proceedings of the National Academy of Sciences*, Vol. 118, No. 26.
27 Haidt, 'Why the past 10 years of American life have been uniquely stupid'.
28 Ibid.

29 Oldenburg, R., 1989, *The Great Good Place* (De Capo Press).
30 Cited in Meltzer, M., 2002, *Mark Twain Himself: A Pictorial Biography* (University of Missouri Press), p. 82.
31 Wirth, L., 1938, 'Urbanism as a way of life', *American Journal of Sociology*, Vol. 44, No. 1.
32 Ibid., pp. 1 and 13.
33 Fischer, C., 1982, *To Dwell Among Friends* (University of Chicago Press).
34 Mumford, L., 1938, *The Culture of Cities* (reprint by Open Road Media, 2016, accessed online at Perlego.com).
35 Putnam, *Bowling Alone*.
36 Woetzel,, J., et al., 2018, 'Smart cities: digital solutions for a more livable future', McKinsey Global Institute (mckinsey.com).
37 Yeung, P., 2022, '"It's a beautiful thing": how one Paris district rediscovered conviviality', *Guardian*.

Chapter 7 – Beyond the Rich World

1 Reza, P., 2016, 'Old photos bring back sweet memories of Bangladesh's capital Dhaka', Global Voices (globalvoices.org).
2 *World Population Review*, 2023, 'Dhaka Population 2023' (worldpopulationreview.com).
3 Jedwab, R. and Vollrath, D., 2015, 'Urbanization without growth in historical perspective', *Explorations in Economic History*, Vol. 58.
4 Authors' calculations based on Oxford Economics data.
5 Bairoch, *Cities and Economic Development*, p. 347.
6 Authors' calculations based on United Nations data.
7 Authors' calculations based on Maddison Project database.
8 Roser, M., Ortiz-Ospina, E. and Ritchie, H. 2019, 'Life expectancy', Our World in Data (ourworldindata.org).
9 United Nations, 1969, 'Growth of the world's urban and rural population, 1920–2000' (un.org).
10 Ritchie and Roser, 'Urbanization'.
11 Victor, H., 1861, *The Sack of the Summer Palace*, Fondation Napoleon (napoleon.org).
12 Jedwab and Vollrath, 'Urbanization without growth in historical perspective'.
13 World Population Review, 2022, 'World city populations 2022' (worldpopulationreview.com).

14 Ibid.
15 For a summary, see Ritchie and Roser, 'Urbanization'.
16 Authors' calculations based on Maddison Project database.
17 Ibid.
18 Jain, 'Case study on territorial development in Japan'.
19 Yokoyama, M., 1991, 'Abortion policy in Japan: analysis from the framework of interest groups', *Kokugakuin Journal of Law and Politics*, Vol. 29, No. 1.
20 Institute for International Cooperation, Japan International Cooperation Agency, 2005, 'Japan's experiences in public health and medical systems' (openjicareport.jica.go.jp).
21 O'Neill, A., 2022, 'Total fertility rate in Japan from 1800 to 2020', Statista (statista.com).
22 Authors' calculations based on Maddison Project database.
23 Kim, K., 1995, 'The Korean miracle (1962–80) revisited: myths and realities in strategy and development', in Stein, H. (ed.), *Asian Industrialization and Africa*, International Political Economy Series (Palgrave Macmillan).
24 Ibid.
25 Joo, Y.-M., 2018, *Megacity Seoul: Urbanization and the Development of Modern South Korea* (Routledge).
26 Ibid.
27 Ibid.
28 World Bank, 2022, 'Four decades of poverty reduction in China' (worldbank.org).
29 Authors' calculations based on Maddison Project database.
30 Whyte, M., et al., 2015, 'Challenging myths about China's one-child policy', *The China Journal*, Vol. 74.
31 Sheng, A. and Geng, X., 2018, 'China is building 19 "supercity clusters"', World Economic Forum (weforum.org).
32 Preen, M., 2018, 'China's city clusters: the plan to develop 19 super-regions', China Briefing (china-briefing.com).
33 *The Economist*, 2022, 'Xi Jinping's economic revolution aims to spread growth'.
34 Authors' calculations based on World Bank data.
35 Jedwab and Vollrath, 'Urbanization without growth in historical perspective'.
36 Griffith, J., 2019, '22 of the top 30 most polluted cities in the world are in India', CNN (cnn.com).

37 World Bank Data, 2018, 'Population living in slums (% of urban population)' (data.worldbank.org).

38 Nijman, J. and Shin, M., 2014, 'The megacity', in Knox, P. (ed.), *Atlas of Cities* (Princeton University Press).

39 Ibid.

40 Marx, B., et al., 2013, 'The economics of slums in the developing world', *Journal of Economic Perspectives*, Vol. 27, No. 4.

41 Glaeser, E., 2011, *Triumph of the City* (Penguin).

42 Ibid.

43 International Labour Organization, 2018, 'Women and men in the informal economy: a statistical picture' (ilo.org).

44 WIEGO Statistics Database (wiego.org).

45 Samora, P., 2016, 'Is this the end of slum upgrading in Brazil?', The Conversation (theconversation.com).

46 Marx, et al., 'The economics of slums in the developing world'.

47 Farouk, M., 2020, 'Cautious hopes for slum dwellers relocated in Egypt housing project', Reuters.

48 *Daily News Egypt*, 2021, 'Egypt to be free from informal housing areas by 2030: Cabinet'.

49 Farouk, 'Cautious hopes for slum dwellers relocated in Egypt housing project'.

50 de Soto, H., 2000, *The Mystery of Capital* (Basic Books).

51 Marx, et al., 'The economics of slums in the developing world'.

52 Martine, G., et al., 2013, 'Urbanization and fertility decline: cashing in on structural change', IIED Human Settlements working paper.

53 Chandna, H., 2018, 'The Rise of the Tycoon', *Business World*.

54 Mahurkar, U., 2011, 'A look into profile of billionaire Gautam Adani', *India Today*.

55 Roser, M., 2017, 'Fertility rate', Our World in Data (ourworldindata .org).

56 O'Neill, 'Total fertility rate in India, from 1880 to 2020'.

57 Roser, M., 'Fertility rate'.

58 Irwin, D., 2020, 'The rise and fall of import substitution', Peterson Institute of International Economics (piie.com).

59 Little, I., et al., 1970, *Industry and Trade in Some Developing Countries* (OECD and Oxford University Press).

60 Ritchie, H. and Roser, M., 2021, 'Crop yields', Our World in Data (ourworldindata.org).

61 Hoeffler, A., et al., 2011, 'Post-conflict recovery and peacebuilding', World Bank (worldbank.org).
62 Gollin, D., et al., 2016, 'Urbanization with and without industrialization', *Journal of Economic Growth*, Vol. 21.
63 World Bank Data, 2022, 'Research and development expenditure (% of GDP)' (data.worldbank.org).
64 Rodrik, D., 2022, 'Prospects for global economic convergence under new technologies', in Autor, D., et al. (eds), *An Inclusive Future? Technology, New Dynamics, and Policy Challenges*, Brookings Institution (brookings.edu).
65 Andersson, J., et al., 2018, 'Is apparel manufacturing coming home?', McKinsey & Company (mckinsey.com).
66 Startup Genome, 2022, 'Startup ecosystems' (startupgenome.com).
67 Ibid.

Chapter 8 – The Spectre of Disease

1 Jedwab, R., et al., 2019, 'Pandemics, places, and populations: evidence from the Black Death', CESifo Working Paper.
2 Roos, D., 2020, 'Social distancing and quarantine were used in medieval times to fight the Black Death', History.com.
3 Jedwab, et al., 'Pandemics, places, and populations'.
4 Eyewitness to History, 'The Black Death, 1348' (eyewitnesstohistory .com).
5 Bray, R., 2020, *Armies of Pestilence: The Impact of Pandemics on History* (Lutterworth Press).
6 Ibid.
7 Ibid.
8 Ibid.
9 Kenny, C., 2021, *The Plague Cycle* (Simon & Schuster).
10 Greger, M., 2007, 'The human/animal interface: emergence and resurgence of zoonotic infectious diseases', *Critical Review in Biology*, Vol. 33, No. 4.
11 *The Economist*, 2021, 'The pandemic's true death toll'.
12 For a full discussion of the impact of the Covid-19 pandemic, see Goldin, I., 2022, *Rescue: From Global Crisis to a Better World* (Sceptre).
13 Kharas, H. and Dooley, M., 2021, 'Long-run impacts of Covid-19 on extreme poverty', Brookings Institution (brookings.edu).

14 Sánchez-Páramo, C., et al., 2021, 'Covid-19 leaves a legacy of rising poverty and widening inequality', World Bank (worldbank.org).
15 Kharas and Dooley, 'Long-run impacts of Covid-19 on extreme poverty'.
16 UN Women, 2022, Ukraine and the food and fuel crisis: 4 things to know, unwomen.org.
17 Hayward, E., 2021, 'Covid-19's toll on mental health', Boston College (bc.edu).
18 *The Economist*, 2022, 'Covid learning loss has been a global disaster'.
19 World Bank, UNESCO and UNICEF, 2021, 'The state of the global education crisis: a path to recovery' (worldbank.org).
20 *The Economist*, 'Covid learning loss has been a global disaster'.
21 Ibid.
22 Harper, K. and Armelagos, G., 2010, 'The changing disease-scape in the third epidemiological transition', *International Journal of Environmental Research and Public Health*, Vol. 7, No. 2.
23 Kenny, *The Plague Cycle*, p. 177.
24 World Bank Data, 2022, 'Air transport, passengers carried' (data .worldbank.org).
25 Davenport, R., 2020, 'Urbanization and mortality in Britain, c. 1800–50', *The Economic History Review*, Vol. 73, No. 2.
26 O'Neill, A., 2022, 'Child mortality rate (under five years old) in the United Kingdom from 1800 to 2020', Statista (statista.com).
27 Kenny, *The Plague Cycle*.
28 Ibid.
29 Ibid.
30 Jones, K., et al., 2008, 'Global trends in emerging infectious diseases', *Nature*, Vol. 451.
31 Morand, S. and Walther, B., 2020, 'The accelerated infectious disease risk in the Anthropocene: more outbreaks and wider global spread', Working paper.
32 Centers for Disease Control and Prevention, 2021, 'Zoonoses' (cdc .gov).
33 Greger, 'The human/animal interface'.
34 Ibid.
35 UNFAO, 2022, 'State of the world's forests' (fao.org).
36 Young, R., 2014, 'Take bushmeat off the menu before humans are served another Ebola', The Conversation (theconversation.com).

37 Cawthorn, D-M. and Hoffman, L., 2015, The bushmeat and food security nexus: A global account of the contributions, conundrums and ethical collisions, *Food Research International*, Vol. 76.
38 Maxmen, A., 2022, 'Wuhan market was epicentre of pandemic's start, studies suggest', Nature.com.
39 Kenny, *The Plague Cycle*.
40 Smitham, E. and Glassman, A., 2021, 'The next pandemic could come sooner and be deadlier', Center for Global Development (cgdev.org).
41 Barnes, O., 2022, 'Just where and when will the next pandemic strike?', *Financial Times*.
42 Ibid.
43 Kenny, *The Plague Cycle*.
44 Discovery's Edge, 2021, 'Researchers clarify why measles doesn't evolve to escape immunity' (discoverysedge.mayo.edu).
45 Lobanovska, M. and Pilla, G., 2017, 'Penicillin's discovery and antibiotic resistance: lessons for the future?', *Yale Journal of Biology and Medicine*, Vol. 90, No. 1.
46 The Week, 2021, 'The rise of the superbugs: why antibiotic resistance is a "slow-moving pandemic"' (theweek.co.uk).
47 Ibid.
48 Murray, C., et al., 2022, 'Global burden of bacterial antimicrobial resistance in 2019: a systematic analysis', *The Lancet*, Vol. 399, No. 10325.
49 Sample, I., 2007, 'Drug-resistant form of plague identified', *Guardian*.
50 Pavel, B. and Venkatram, V., 2021, 'Facing the future of bioterrorism', Atlantic Council (atlanticcouncil.org).
51 Langer, R. and Sharma, S., 2020, 'The blessing and curse of biotechnology: a primer on biosafety and biosecurity', Carnegie Endowment for International Peace (carnegieendowment.org).
52 Ibid.
53 Stier, A., et al., 2020, 'COVID-19 attack rate increases with city size', Working paper.
54 Bentley, A., et al., 2018, 'Recent origin and evolution of obesity–income correlation across the United States', *Palgrave Communications*.
55 Authors' calculations based on data from CDC and NYC Health.
56 Authors' calculations based on data from CDC.
57 NYC Health, 2022, 'Covid-19: data' (nyc.gov).

58 Citi GPS, 2020, 'Technology at Work v5.0: A new world of remote work' (oxfordmartin.ox.ac.uk).
59 Wade, L., 2020, 'From Black Death to fatal flu, past pandemics show why people on the margins suffer most', Science.org.
60 Authors' calculations based on *The Economist*'s Excess Deaths model.
61 Ibid.
62 Goldin, *Rescue*.
63 Gates, B., 2022, *How to Prevent the Next Pandemic* (Allen Lane).
64 Jedwab, et al., 'Pandemics, places, and populations'.

Chapter 9 – A Climate of Peril

1 Gabour, J., 2015, 'A Katrina survivor's tale: "They forgot us and that's when things started to get bad"', *Guardian*.
2 Heggie, J., 2018, 'Day Zero: Where next?', *National Geographic*.
3 Gladstone, N., 2019, 'Air quality ten times hazardous levels in Sydney', *Sydney Morning Herald*.
4 Goldin and Muggah, *Terra Incognita*, p. 67.
5 Mathis, W. and Kar, T., 2022, 'UK temperatures reach highest on record, Met Office says', Bloomberg News.
6 Vaughan, A., 2022, '40°C heatwave may have killed 1000 people in England and Wales', *New Scientist*.
7 Mallapaty, S., 2022, 'Why are Pakistan's floods so extreme this year?', Nature.com.
8 Colbert, A., 2022, 'A force of nature: hurricanes in a changing climate', NASA (climate.nasa.gov).
9 Carpineti, A., 2022, 'Nine of the top 10 hottest years ever all occurred in the last decade', IFL Science (iflscience.com).
10 Boland, B., et al., 2021, 'How cities can adapt to climate change', McKinsey & Company (mckinsey.com).
11 Urban Climate Change Research Network, 2018, 'The Future we don't want' (c40.org).
12 Lieberman, B. and Gordon, E., 2021 (2nd edition), *Climate Change in Human History* (Bloomsbury).
13 Ibid.
14 Ibid.
15 Haug, G., et al., 2001, 'Southward migration of the intertropical convergence zone through the Holocene', *Science*, Vol. 293, No. 5533.

16 Lieberman and Gordon, *Climate Change in Human History*.
17 Cline, E., 2014, *1177 B.C.: The Year Civilization Collapsed* (Princeton University Press).
18 Ibid.
19 Holder, J., et al., 2017, 'The three-degree world: the cities that will be drowned by global warming', *Guardian*.
20 Ibid.
21 Ibid.
22 Kimmelman, M., 2017, 'The Dutch have solutions to rising seas. The world is watching', *New York Times*.
23 Ibid.
24 Hussain, Z., 2022, 'Why Indonesia's Jakarta is the fastest sinking city in the world', *India Times*.
25 Tarigan, E. and Karmini, N., 2022, 'Indonesia's sinking, polluted capital is moving to new city', Associated Press.
26 European Commission Joint Research Centre, 2022, 'Cities are often 10–15 °C hotter than their rural surroundings' (joint-research-centre.ec.europa.eu).
27 Lieberman and Gordon, *Climate Change in Human History*.
28 Goldin and Muggah, *Terra Incognita*, pp. 68–9.
29 Arora, N., 2021, 'Air pollution led to around 54,000 premature deaths in New Delhi in 2020 – study', Reuters.
30 Abraham, B., 2022, 'Deadly heatwave also resulting in ozone exceedance and it's making Delhi's air toxic', *India Times*.
31 Jordan, J., 2018, 'Living on the frontiers of climate change: a mother's story', YouTube (youtube.com).
32 Goldin and Muggah, *Terra Incognita*.
33 Wester, P., et al., 2019, *The Hindu Kush Himalaya Assessment* (Springer).
34 Goldin and Muggah, *Terra Incognita*, p. 53.
35 Bodewig, C., 2019, 'Climate change in the Sahel: how can cash transfers help protect the poor?', Brookings Institution (brookings.edu).
36 Ibid.
37 Al Jazeera, 2011, 'Kashmir and the politics of water' (aljazeera.com).
38 Moran, D., et al., 2018, 'Carbon footprints of 13 000 cities', *Environmental Research Letters*, Vol. 13, No. 6.
39 Swinney, P., 2019, 'Are cities bad for the environment?', BBC (bbc.com).

40 Jones, C. and Kammen, D., 2013, 'Spatial distribution of U.S. household carbon footprints reveals suburbanization undermines greenhouse gas benefits of urban population density', *Environmental Science & Technology*, Vol. 48, No. 2.
41 Ibid.
42 Glaeser, E. and Kahn, M., 2010, 'The greenness of cities: carbon dioxide emissions and urban development', *Journal of Urban Economics,* Vol. 67, No. 3.
43 CoolClimate Network, 2013, 'Household calculator' (coolclimate .berkley.edu).
44 Ritchie, H., 2020, 'Which form of transport has the smallest carbon footprint?', Our World in Data (ourworldindata.org).
45 Freemark, Y., 2018, 'Travel mode shares in the U.S.', The Transport Politic (thetransportpolitic.com).
46 Ibid.
47 Sustain Europe, 2019, 'Oslo European Green Capital 2019' (sustaineurope.com).
48 Norwegian Ministry of Petroleum and Energy, 2016, 'Renewable energy production in Norway' (regjeringen.no).
49 Ferenczi, A., 2021, 'A city without cars is already here, and it's idyllic', Vice (vice.com).
50 Wilson, E. O., 1984, *Biophilia* (Harvard University Press).
51 Montgomery, C., 2013, *Happy City* (Penguin).
52 Trust for Public Land, 2022, 'ParkScore index' (tpl.org).
53 Ibid.
54 Singapore National Parks, 2020, 'Annual report' (nparks.gov.sg).
55 Singapore National Parks, 2020, 'A city in nature' (nparks.gov.sg).
56 O'Sullivan, F., 2019, 'Paris wants to grow "urban forests" at famous landmarks', Bloomberg News.
57 Kennedy, C., et al., 2015, 'Energy and material flows of megacities', *Proceedings of the National Academy of Sciences*, Vol. 112, No. 19.
58 Ritchie, H., 2020, 'Food waste is responsible for 6% of global greenhouse gas emissions', Our World in Data (ourworldindata .org).
59 Oakes, K., 2020, 'How cutting your food waste can help the climate', BBC (bbc.com).
60 Biba, E., 2019, 'Everything Americans think they know about recycling is probably wrong', NBC (nbcnews.com).

61 Environmental Protection Agency, 2022, 'National overview: facts and figures on materials, wastes and recycling' (epa.gov).
62 Ibid.
63 Villazon, L., 2022, 'Vertical farming: why stacking crops high could be the future of agriculture', Science Focus (sciencefocus.com).
64 Maglio, N., 2022, 'Green forges, how different types of agriculture impact CO2 emissions', Green Forges (greenforges.com).
65 Ritchie, H., 2019, 'Who has contributed most to global CO_2 emissions?', Our World in Data (ourworldindata.org).
66 Hausfather, Z., 2017, 'Mapped: the world's largest CO2 importers and exporters', Carbon Brief (carbonbrief.org).
67 Ibid.
68 Postdam Institute for Climate Impact Research, 2022, 'Tipping elements – the Achilles' heels of the earth system' (pik-postdam.de).
69 Ibid.

Chapter 10 – Conclusion: Better Together

1 Mumford, *The City in History*, p. 3.
2 UN Department of Economic and Social Affairs, Population Division, 2022, 'World population prospects' (population.un.org).
3 OECD, 2015, *The Metropolitan Century* (oecd.org).
4 Hess, D. and Rehler, J., 2021, 'America has eight parking spaces for every car. Here's how cities are rethinking that land', *Fast Company*.
5 Morris, D., 2016, 'Today's cars are parked 95% of the time', *Fortune*.
6 Schneider, B., 2018, 'CityLab University: induced demand', Bloomberg News.
7 Jacobs, J., 1961, *The Death and Life of Great American Cities* (Random House).
8 Rushton, D., 2020, 'London vs New York: which city has the higher average building height?', Carto (carto.com).
9 Montgomery, C., 2013, *The Happy City* (Penguin Books).
10 OECD Data, 2021, 'Net ODA' (data.oecd.org).
11 Goldin and Muggah, *Terra Incognita*, p. 133.
12 Thatcher Archive, Interview for *Woman's Own*, 23 September 1987, Margaret Thatcher Foundation (margaretthatcher.org).

BIBLIOGRAPHY

Abel, J. and Deitz, R., 2019, 'Why are some places so much more unequal than others?', *Economic Policy Review*, Vol. 25, No. 1.

Abraham, B., 2022, 'Deadly heatwave also resulting in ozone exceedance and it's making Delhi's air toxic', *India Times*.

Acemoglu, D. and Restrepo, P., 2019, 'Automation and new tasks: how technology displaces and reinstates labor', *Journal of Economic Perspectives*, Vol. 33, No. 2.

Al Jazeera, 2011, 'Kashmir and the politics of water' (aljazeera.com).

Al-Khalili, J., 2012, *The House of Wisdom* (Penguin).

Allen, T., 1977, *Managing the Flow of Technology* (MIT Press).

Anderson, B., 1983, *Imagined Communities* (Verso).

Anderson, D., 2021, 'Home prices in cities rise 16%, surpassing suburban and rural price growth for the first time since pre-pandemic', Redfin (redfin.com).

Andersson, J., et al., 2018, 'Is apparel manufacturing coming home?', McKinsey & Company (mckinsey.com).

Arora, N., 2021, 'Air pollution led to around 54,000 premature deaths in New Delhi in 2020 – study', Reuters.

Augustine, 426, *The City of God: Volume I* (reprint by Jazzybee Verlag, 2012, accessed online at Perlego.com).

Autor, D., 2019, 'Work of the past, work of the future', *AEA Papers and Proceedings*, Vol. 109.

Autor, D., et al., 2017, 'Gentrification and the amenity value of crime reductions', NBER working paper.

Ayers, D., 2019, 'The gender gap in marriages between college-educated partners', Institute for Family Studies (ifstudies.org).

Bairoch, P., 1988, *Cities and Economic Development* (University of Chicago Press).

Ball, J., 2019, 'Housing as a basic human right', *The New Statesman*.

Barjamovic, G., et al., 2019, 'Trade, merchants, and the lost cities of the Bronze Age', *The Quarterly Journal of Economics*, Vol. 134, No. 3.

Barnes, O., 2022, 'Just where and when will the next pandemic strike?', *Financial Times*.

Bekhrad, J., 2017, 'The obscure religion that shaped the West', BBC (bbc.com).

Bellah, R., 2011, *Religion in Human Evolution* (Harvard University Press).

Bentley, A., et al., 2018, 'Recent origin and evolution of obesity–income correlation across the United States', *Palgrave Communications*.

Berkes, E. and Gaetani, R., 2021, 'The geography of unconventional innovation', *The Economic Journal*, Vol. 131, No. 636.

Berube, A. and Parilla, J., 2012, 'Metro trade: cities return to their roots in the global economy', Brookings Institution (brookings.edu).

Biba, E., 2019, 'Everything Americans think they know about recycling is probably wrong', NBC (nbcnews.com).

Bloom, N., Twitter, 29 August 2022, 14:59.

Bloom, N., et al., 2015, 'Does working from home work? Evidence from a Chinese experiment', *The Quarterly Journal of Economics*, Vol. 130, No. 1.

Bodewig, C., 2019, 'Climate change in the Sahel: how can cash transfers help protect the poor?', Brookings Institution (brookings.edu).

Bonnet, F., 2014, 'Social housing in New York', Metropolitics (metropolitics.org).

Bor, A. and Bang Petersen, M., 2021, 'The psychology of online political hostility: a comprehensive, cross-national test of the mismatch hypothesis', *American Political Sciences Review*, Vol. 116, No. 1.

Boustan, L., et al., 2013, 'Urbanization in the United States, 1800–2000', NBER working paper.

Bray, R., 2020, *Armies of Pestilence: The Impact of Pandemics on History* (Lutterworth Press).

Brooks, D., 2000, *Bobos in Paradise* (Simon & Schuster).

Broom, D., 2021, 'Home or office? Survey shows opinions about work after COVID-19', World Economic Forum (weforum.org).

C40 Cities, 2018, 'The Future We Don't Want' (c40.org).

Cadman, E., 2021, 'Your new office set-up is going to look a lot like the old one', Bloomberg News.

Cairncross, F., 2001, *The Death of Distance* (Harvard Business Review Press).

Cantor, N. F., 2002, *In the Wake of the Plague: The Black Death and the World It Made* (HarperCollins).

Carpineti, A., 2022, 'Nine of the top 10 hottest years ever all occurred in the last decade', IFL Science (iflscience.com).

Cawthorn, D-M. and Hoffman, L., 2015, The bushmeat and food security nexus: A global account of the contributions, conundrums and ethical collisions, *Food Research International*, Vol. 76

Centers for Disease Control and Prevention, 2021, 'Zoonoses' (cdc.gov).

Chandler, T., and Fox, G., 1974, *3000 Years of Urban Growth* (Academic Press, accessed online at Perlego.com).

Chandna, H., 2018, 'The rise of the tycoon', *Business World*.

Chen, W., 2013, 'Internet use, online communication, and ties in Americans' networks', *Social Science Computer Review*, Vol. 31, No. 4.

Chernick, H., et al., 2011, 'Revenue diversification in large U.S. cities', IMFG Papers on Municipal Finance and Governance.

Cho, J., et al., 2020, 'Do search algorithms endanger democracy? An experimental investigation of algorithm effects on political polarization', *Journal of Broadcasting & Electronic Media*, Vol. 64, No. 2.

Cintu, A., 2021, 'Record apartment conversions make 2021 most successful year in adaptive reuse', RentCafe (rentcafe.com).

Citi GPS, 2020, 'Technology at work v5.0: a new world of remote work' (oxfordmartin.ox.ac.uk).

Clark, D., 2022, 'Average age of mothers at childbirth in the United Kingdom from 1938 to 2020', Statista (statista.com).

Clark, W., 'School desegregation and white flight: a re-examination and case study', *Social Science Research*, Vol. 16, No. 3.

Cline, E., 2014, *1177 B.C.: The Year Civilization Collapsed* (Princeton University Press).

Clossick, J., 2014, 'The industrial city', in Knox, P. (ed.), *Atlas of Cities* (Princeton University Press).

Cohen, B., 2022, 'Yes, Zoom has an office. No, it's not a place to work', *Wall Street Journal*.

Colbert, A., 2022, 'A force of nature: hurricanes in a changing climate', NASA (climate.nasa.gov).

CoolClimate Network, 2013, 'Household calculator' (coolclimate .berkley.edu).

Cross, R., et al., 2021, 'Collaboration overload is sinking productivity', *Harvard Business Review*.

D'Onfro, J., 2015, 'Steve Jobs had a crazy idea for Pixar's office to force people to talk more', *Business Insider*.

Daily News Egypt, 2021, 'Egypt to be free from informal housing areas by 2030: Cabinet' (dailynewsegypt.com).

Davenport, R., 2020, 'Urbanization and mortality in Britain, c. 1800–50', *The Economic History Review*, Vol. 73, No. 2.

de Blij, H., 2008, *The Power of Place* (Oxford University Press).

de Soto, H., 2000, *The Mystery of Capital* (Basic Books).

de Tocqueville, A., 1958 (written 1833–5), *Journeys to England and Ireland* (reprint by Routledge, 2017).

de Vise, Pierre, 1976, 'The suburbanization of jobs and minority employment', *Economic Geography*, Vol. 52, No. 4.

The Dearborn Independent, 1922, *Ford Ideals* (Dearborn Publishing Company).

Deibert, R., 1997, *Parchment, Printing, and Hypermedia: Communication in World Order Transformation* (Columbia University Press).

DeVoe, J., 2021, 'Leaders say Arlington has the ecosystem to be the next Austin or Miami', ARL Now.

Diamond, R. and Moretti, E., 2021, 'Where is standard of living the highest? Local prices and the geography of consumption', NBER working paper.

Discovery's Edge, 2021, 'Researchers clarify why measles doesn't evolve to escape immunity' (discoverysedge.mayo.edu).

Dougherty, C. and Burton, A., 2017, 'A 2:15 alarm, 2 trains and a bus get her to work by 7 a.m.', *New York Times*.

Dunbar, K., 1997, 'How scientists think: on-line creativity and conceptual change in science', in Ward, T. B., et al. (eds), *Conceptual Structures and Processes: Emergence, Discovery, and Change* (American Psychological Association Press).

Dunbar, R. 1992, 'Neocortex size as a constraint on group size in primates', *Journal of Human Evolution*, Vol. 22, No. 6.

Easterly, W., et al., 2016, 'A long history of a short block: four centuries of development surprises on a single stretch of a New York city street' (greenestreet.nyc/paper).
The Economist, 2020, 'Governments are rethinking the provision of public housing'.
The Economist, 2021, 'The pandemic's true death toll'.
The Economist, 2021, 'The rise of working from home'.
The Economist, 2022, 'Covid learning loss has been a global disaster'.
The Economist, 2022, 'The future of public transport in Britain'.
The Economist, 2022, 'Xi Jinping's economic revolution aims to spread growth'.
Edlund, L., et al., 'Gentrification and the rising returns to skill', NBER working paper.
Ehrenhalt, A., 2013, *The Great Inversion* (Vintage).
Elliott, L., 2019, 'HS2 would widen UK north–south divide and should be axed, says report', *Guardian*.
Engels, F., 1845, *The Condition of the Working Class in England* (reprint by Penguin Classics, 2009).
Environmental Protection Agency, 2022, 'National Overview: Facts and Figures on Materials, Wastes and Recycling' (epa.gov).
European Commission Joint Research Centre, 2022, 'Cities are often 10–15 °C hotter than their rural surroundings' (joint-research-centre.ec.europa.eu).
Eyewitness to History, 'The Black Death, 1348' (eyewitnesstohistory.com).
Farouk, M., 2020, 'Cautious hopes for slum dwellers relocated in Egypt housing project', Reuters.
Federal Reserve Bank of New York, 2021, 'The labor market for recent college graduates' (newyorkfed.org).
Federal Reserve Bank of St Louis, 2019, 'The US auto labor market since NAFTA' (stlouisfed.org).
Feldman, E., 2022 'Converting office buildings to apartments could help city housing crisis', Spectrum News.
Ferenczi, A., 2021, 'A city without cars is already here, and it's idyllic', Vice (vice.com).
Ferguson, N., 2017, *The Square and the Tower* (Penguin).
Fessenden, M., 2015, 'Making a sandwich from scratch took this man six months', *Smithsonian Magazine*.

Fischer, C., 1982, *To Dwell Among Friends* (University of Chicago Press).

—, 2007, 'Technology and society: historical complexities', *Sociological Inquiry*, Vol. 67, No. 1.

Florida, R., 2002, *The Rise of the Creative Class* (Basic Books).

Florida, R., et al., 2008, 'The rise of the mega-region', *Cambridge Journal of Regions, Economy and Society*, Vol. 1, No. 3.

Frankopan, P., 2015, *The Silk Roads* (Bloomsbury).

Freemark, Y., 2018, 'Travel mode shares in the U.S.', The Transport Politic (thetransportpolitic.com).

Friedman, T., 2007, *The World is Flat* (Penguin).

Frost, R., 2020, 'Are Americans stuck in place? Declining residential mobility in the US', Joint Center for Housing Studies of Harvard University (jchs.harvard.edu).

Gabour, J., 2015, 'A Katrina survivor's tale: "They forgot us and that's when things started to get bad"', *Guardian*.

Gates, B., 2022, *How to Prevent the Next Pandemic* (Allen Lane).

George, A., 1999, *The Epic of Gilgamesh* (Penguin Classics).

Gibbs, M., et al., 2021, 'Work from home & productivity: evidence from personnel & analytics data on IT professionals', *SSRN Electronic Journal*.

Gibson, C., 1998, 'Population of the 100 largest cities and other urban places in the United States: 1790 to 1990', US Census Bureau (census.gov).

Gladstone, N., 2019, 'Air quality ten times hazardous levels in Sydney', *Sydney Morning Herald*.

Glaeser, E., 2005, 'Reinventing Boston: 1630–2003', *Journal of Economic Geography*, Vol. 5, No. 2.

—, 2011, *Triumph of the City* (Penguin).

Glaeser, E. and Ponzetto, G., 2007, 'Did the death of distance hurt Detroit and help New York?', NBER working paper.

Glaeser, E. and Kahn, M., 2010, 'The greenness of cities: carbon dioxide emissions and urban development', *Journal of Urban Economics*, Vol. 67, No. 3.

Glass, R., 1964, *London: Aspects of Change* (MacGibbon & Kee).

Goldin, I., 2022, *Rescue: From Global Crisis to a Better World* (Sceptre).

Goldin, I. and Muggah, R., 2020, *Terra Incognita: 100 Maps to Survive the Next 100 Years* (Century).

Goldin, I., et al., 2022, 'Why is productivity slowing down', *Journal of Economic Literature* (forthcoming).

Goldstein, A., 2017, *Janesville: An American Story* (Simon & Schuster).

Gollin, D., et al., 2016, 'Urbanization with and without industrialization', *Journal of Economic Growth*, Vol. 21.

Goos, M. and Manning, A., 2007, 'Lousy and lovely jobs: the rising polarization of work in Britain', *The Review of Economics and Statistics*, Vol. 89. No. 1.

Greger, M., 2007, 'The human/animal interface: emergence and resurgence of zoonotic infectious diseases', *Critical Review in Biology*, Vol. 33, No. 4.

Griffith, J., 2019, '22 of the top 30 most polluted cities in the world are in India', CNN (cnn.com).

Grimbly, S., 2013, *Encyclopedia of the Ancient World* (Fitzroy Dearborn).

Gropius, W., 1969, 'Principles of Bauhaus production', in Wingler, H. (ed.), *The Bauhaus* (MIT Press).

The *Guardian* Data Blog, 2016, 'Deprivation and poverty in London', *Guardian*.

Gwendolyn, L., 2001, *Mesopotamia: The Invention of the City* (Penguin).

Haidt, J., 2022, 'Why the past 10 years of American life have been uniquely stupid', *The Atlantic*.

Hall, P., 1998, *Cities in Civilization* (Weidenfeld & Nicolson).

Hammond, G., 2021, 'Canary Wharf: does the east London office district have a future?', *Financial Times*.

Hannah-Jones, N., 2019, 'It was never about busing', *New York Times*.

Harari, Y. N., 2014, *Sapiens* (Harper).

Harmon, J. and Zim, E., 2021, 'The rise and fall of the American mall', *Business Insider*.

Harper, K. and Armelagos, G., 2010, 'The changing disease-scape in the third epidemiological transition', *International Journal of Environmental Research and Public Health*, Vol. 7, No. 2.

Hartman, A., 2021, 'Salesforce says "the 9-to-5 workday is dead" and will provide 3 new ways for employees to work – including the possibility of working from home forever', *Business Insider*.

Haug, G., et al., 2001, 'Southward migration of the intertropical convergence zone through the Holocene', *Science*, Vol. 293, No. 5533.

Hausfather, Z., 2017, 'Mapped: the world's largest CO2 importers and exporters', Carbon Brief (carbonbrief.org).

Haverfield, F., 1904, *The Romanization of Roman Britain* (reprint by Perlego, 2004, accessed at Perlego.com).

Havsy, J., 2022, 'Why office-to-apartment conversions are likely a fringe trend at best', Moody's Analytics (cre.moodysanalytics.com).

Hayward, E., 2021, 'COVID-19's toll on mental health', Boston College (bc.edu).

Hazlitt, W., 1837, *Characteristics in the manner of Rochefoucault's Maxims* (J. Templeman).

Heggie, J., 2018, 'Day Zero: where next?', *National Geographic.*

Hess, D. and Rehler, J., 2021, 'America has eight parking spaces for every car. Here's how cities are rethinking that land', *Fast Company.*

Hobbs, F. and Stoops, N., 2002, 'Demographic trends in the 20th century', US Census Bureau (census.gov).

Hoeffler, A., et al., 2011, 'Post-conflict recovery and peacebuilding', World Bank (worldbank.org).

Howard, E., 1898, *Garden Cities of Tomorrow* (reprint by Perlego, 2014, accessed online at Perlego.com).

Huber, M., 2013, *Lifeblood: Oil, Freedom, and the Forces of Capital* (University of Minnesota Press).

Hugo, V., 1861, 'The sack of the Summer Palace', Fondation Napoleon (napoleon.org).

Hussain, Z., 2022, 'Why Indonesia's Jakarta is the fastest sinking city in the world', *India Times*.

Institute for Economics and Peace, 2020, 'Ecological Threat Register' (ecologicalthreatregister.org).

Institute for International Cooperation, Japan International Cooperation Agency, 2005, 'Japan's experiences in public health and medical systems' (openjicareport.jica.go.jp).

International Labour Organization, 2018, 'Women and men in the informal economy: a statistical picture' (ilo.org).

Irwin, D., 2020, 'The rise and fall of import substitution', Peterson Institute of International Economics (piie.com).

Jacobs, J., 1961, *The Death and Life of Great American Cities* (Random House).

Jaffe, A. B., et al., 1993, 'Geographic localization of knowledge spillovers as evidenced by patent citations', *The Quarterly Journal of Economics*, Vol. 108, No. 3.

Jain, V., 2019, 'Case study on territorial development in Japan', World Bank (worldbank.org).

Jedwab, R., et al., 2019, 'Pandemics, places, and populations: evidence from the Black Death', CESifo Working Paper.

Jedwab, R. and Vollrath, D., 2015, 'Urbanization without growth in historical perspective', *Explorations in Economic History*, Vol. 58.

Jefferson, T., 1800, 'From Thomas Jefferson to Benjamin Rush', National Archives (archives.gov).

Jeunesse, E., 2017, 'The rise of poverty in suburban and outlying areas', Joint Center for Housing Studies (jchs.harvard.edu).

Johnson, S., 2010, *Where Good Ideas Come From* (Penguin).

Jones, C. and Kammen, D., 2013, 'Spatial distribution of U.S. household carbon footprints reveals suburbanization undermines greenhouse gas benefits of urban population density', *Environmental Science & Technology*, Vol. 48, No. 2.

Jones, K., et al., 2008, 'Global trends in emerging infectious diseases', *Nature*, Vol. 451.

Joo, Y.-M., 2018, *Megacity Seoul: Urbanization and the Development of Modern South Korea* (Routledge).

Jordan, J., 2018, 'Living on the Frontiers of Climate Change: A Mother's Story', YouTube (youtube.com).

Kastle Systems, 2022, 'Recreation activity vs. office occupancy' (kastle.com).

Kelly, H., Kramer, A. and Warren, A., 2020, *Emerging Trends in Real Estate: United States and Canada 2020*, PWC and Urban Land Institute.

Kennedy, C., et al., 2015, 'Energy and material flows of megacities', *Proceedings of the National Academy of Sciences*, Vol. 112, No. 19.

Kenny, C., 2021, *The Plague Cycle* (Simon & Schuster).

Kerr, S. and Kerr, W., 2018, 'Immigrant entrepreneurship in America: evidence from the survey of business owners 2007 & 2012', NBER working paper.

Kharas, H. and Dooley, M., 2021, 'Long-run impacts of COVID-19 on extreme poverty', Brookings Institution (brookings.edu).

Kim, K., 1995, 'The Korean miracle (1962–80) revisited: myths and realities in strategy and development', in Stein, H. (ed.), *Asian Industrialization and Africa*, International Political Economy Series (Palgrave Macmillan).

Kim, S. and Margo, R., 2003, 'Historical perspectives on US economic geography', NBER working paper.

Kimmelman, M., 2017, 'The Dutch have solutions to rising seas. The world is watching', *New York Times*.

Kober, E., 2020, 'New York City records another year of jobs–housing mismatch', Manhattan Institute (economics21.org).

Kuhn, D., 2009, *The Age of Confucian Rule* (Harvard University Press).

Langer, R. and Sharma, S., 2020, 'The blessing and curse of biotechnology: a primer on biosafety and biosecurity', Carnegie Endowment for International Peace (carnegieendowment.org).

Lieberman, B. and Gordon, E., 2021 (2nd edition), *Climate Change in Human History* (Bloomsbury).

Lindsey, R. and Dahlman, L., 2022, 'Climate change: global temperature', National Oceanic and Atmospheric Administration (climate.gov).

Little, I., et al., 1970, *Industry and Trade in Some Developing Countries* (OECD and Oxford University Press).

Lobanovska, M. and Pilla, G., 2017, 'Penicillin's discovery and antibiotic resistance: lessons for the future?', *Yale Journal of Biology and Medicine*, Vol. 90, No. 1.

Longman, P., 2015, 'Why the economic fates of America's cities diverged', *The Atlantic*.

Maglio, N., 2022, 'How different types of agriculture impact CO2 emissions', Green Forges (greenforges.com).

Mahurkar, U., 2011, 'A look into profile of billionaire Gautam Adani', *India Today*.

Mallapaty, S., 2022, 'Why are Pakistan's floods so extreme this year?', Nature.com.

Marshall, A., 1890, *Principles of Economics* (reprint by Nabu Press, 2010).

Marshall, T., 2015, *Prisoners of Geography* (Elliott & Thompson).

Martine, G., et al., 2013, 'Urbanization and fertility decline: Cashing in on structural change', IIED Human Settlements Working Paper.

Marvin, C., 1988, *When Old Technologies Were New* (Oxford University Press).

Marx, B., et al., 2013, 'The economics of slums in the developing world', *Journal of Economic Perspectives*, Vol. 27, No. 4.

Mathis, W. and Kar, T., 2022, 'UK temperatures reach highest on record, Met Office says', Bloomberg News.

Matsu, J., et al., 2022, 'Investing in regional equality – lessons from four cities', Chartered Institute of Public Finance and Accountancy (cipfa.org).

Maxmen, A., 2022, 'Wuhan market was epicentre of pandemic's start, studies suggest', Nature.com.

McDonnell, T., 2019, 'Climate change creates a new migration crisis for Bangladesh', *National Geographic*.

McFarland, J., et al., 2018, 'The condition of education 2018', National Centre for Education Statistics (nces.ed.gov).

McLuhan, M., 1962, *The Gutenberg Galaxy* (University of Toronto Press).

Meltzer, M., 2002, *Mark Twain Himself: A Pictorial Biography* (University of Missouri Press).

Meyer, D., 1989, 'Midwestern industrialization and the American manufacturing belt in the nineteenth century', *The Journal of Economic History*, Vol. 49, No. 4.

Miwa, N., 2022, 'High-speed rail development and regional inequalities: evidence from Japan', *Transportation Research Record*.

Mokyr, J., 2016, *Culture of Growth: The Origins of the Modern Economy* (Princeton University Press).

Montgomery, C., 2013, *Happy City* (Penguin).

Moran, D., et al., 2018, 'Carbon footprints of 13 000 cities', *Environmental Research Letters*, Vol. 13, No. 6.

Morand, S. and Walther, B., 2020, 'The accelerated infectious disease risk in the Anthropocene: more outbreaks and wider global spread', Working paper.

Moretti, E., 2010, 'Local multipliers', *American Economic Review*, Vol. 100, No. 2.

—, 2012, *The New Geography of Jobs* (First Mariner Books).

Morgan, K., 2021, 'Why in-person workers may be more likely to get promoted', BBC (bbc.com).

Morphet, J., 2013, 'A city-state?', in Bell, S. and Paskins, J. (eds), *Imagining the Future City: London 2062* (Ubiquity Press).

Morris, D., 2016, 'Today's cars are parked 95% of the time', *Fortune*.

Morris, S. and Hammond, G., 2022, 'Citi plans £100m revamp of Canary Wharf tower', *Financial Times*.

Murray, C., et al., 2022, 'Global burden of bacterial antimicrobial resistance in 2019: a systematic analysis', *The Lancet*, Vol. 399, No. 10325.

NASA (Vital Signs of the Planet), 'Key indicators: Arctic sea ice minimum extent' (climate.nasa.gov).

Nijman, J. and Shin, M., 2014, 'The megacity', in Knox, P. (ed.), *Atlas of Cities* (Princeton University Press).

Nilles, J., 1976, *The Telecommunications–Transportation Tradeoff: Options for Tomorrow* (John Wiley & Sons).

Nordbrandt, M., 2021, 'Affective polarization in the digital age: testing the direction of the relationship between social media and users' feelings for out-group parties', *New Media & Society*.

Norwegian Ministry of Petroleum and Energy, 2016, 'Renewable energy production in Norway' (regjeringen.no).

Norwich, J., 2009, *The Great Cities in History* (Thames & Hudson).

Nuffield Health Report, 2020, 'Working from home taking its toll on the mental health & relationships of the nation' (nuffieldhealth.com).

NYC Health, 2022, 'Covid-19: data' (nyc.gov).

O'Neill, A., 2022, 'Child mortality rate (under five years old) in the United Kingdom from 1800 to 2020', Statista (statista.com).

—, 2022, 'Total fertility rate in Japan from 1800 to 2020', Statista (statista.com).

O'Sullivan, F., 2019, 'Paris wants to grow "urban forests" at famous landmarks', Bloomberg News.

Oakes, K., 2020, 'How cutting your food waste can help the climate', BBC (bbc.com).

OECD Data, 2021, 'Net ODA' (data.oecd.org).

OECD, 2015, 'The metropolitan century' (oecd.org).

Oldenburg, R., 1989, *The Great Good Place* (De Capo Press).

Partridge, J., 2022, 'Google in $1bn deal to buy Central Saint Giles offices in London', *Guardian*.

Paulas, R., 2017, 'The death of the suburban office park and the rise of the suburban poor', *Pacific Standard Magazine*.

Paullin, C. and Wright J., 1932, *Atlas of the Historical Geography of the United States* (reprint by Greenwood Press, 1976).

Pavel, B. and Venkatram, V., 2021, 'Facing the future of bioterrorism', Atlantic Council (atlanticcouncil.org).

Pool, I., 1983, *Forecasting the Telephone* (Praeger).

Postdam Institute for Climate Impact Research, 2022, 'Tipping elements – the Achilles' heels of the earth system' (pik-postdam.de).

Prapotnik, K., et al., 2019, 'Social media as information channel', *Austrian Journal of Political Science*.

Preen, M., 2018, 'China's city clusters: the plan to develop 19 super-regions', China Briefing (china-briefing.com).

Prior, M., 2007, *Post-Broadcast Democracy: How Media Choice Increases Inequality in Political Involvement and Polarizes Elections* (Cambridge University Press).

Puga, D., 2002, 'European regional policies in light of recent location theories', *Journal of Economic Geography*, Vol. 2, No. 4.

Purdy, H., 1945, *A Historical Analysis of the Economic Growth of St Louis, 1840–1945* (Federal Reserve Bank of St Louis).

Putnam, R., 2000, *Bowling Alone* (Simon & Schuster).

Putzier, K., 2021, 'Google to buy New York City office building for $2.1 billion', *Wall Street Journal*.

PWC, 2020, 'Emerging trends in real estate' (pwc.com).

Rajan, R., 2019, *The Third Pillar* (William Collins).

Rathje, S., et al., 2020, 'Out-group animosity drives engagement on social media', *Proceedings of the National Academy of Sciences*, Vol. 118, No. 26.

RATP, 2022, 'Tickets and fares' (ratp.fr).

Reagor, C., 2017, 'Downtown Phoenix's rebirth has been decades in the making. Here's how they did it', *Arizona Republic* (azcentral.com).

Reza, P., 2016, 'Old photos bring back sweet memories of Bangladesh's capital Dhaka', Global Voices (globalvoices.org).

Rhodes, J. and Brown, L., 2019, 'The rise and fall of the "inner city": race, space, and urban policy in postwar England', *Journal of Ethnic and Migration Studies*, Vol. 45, No. 17.

Ritchie, H., 2019, 'Who has contributed most to global CO_2 emissions?', Our World in Data (ourworldindata.org).

—, 2020, 'Food waste is responsible for 6% of global greenhouse gas emissions', Our World in Data (ourworldindata.org).

—, 2020, 'Which form of transport has the smallest carbon footprint?', Our World in Data (ourworldindata.org).

Ritchie, H. and Roser, M., 2019, 'Land use', Our World in Data (ourworldindata.org).

—, 2019, 'Urbanization', Our World in Data (ourworldindata.org).

—, 2021, 'Crop yields', Our World in Data (ourworldindata.org).

Rodrik D., 2022, 'Prospects for global economic convergence under new technologies', in Autor, D., et al. (eds), *An Inclusive Future? Technology, New Dynamics, and Policy Challenges*, Brookings Institution (brookings.edu).

Rogers, T. N., 2022, 'Expensive commutes and $14 lunches: how inflation hampers back-to-work push', *Financial Times*.

Roos, D., 2020, 'Social distancing and quarantine were used in medieval times to fight the Black Death' (History.com).

Roser, M., 2017, 'Fertility rate', Our World in Data (ourworldindata.org).

Roser, M., Ortiz-Ospina, E. and Ritchie, H., 2019, 'Life expectancy', Our World in Data (ourworldindata.org).

Roser, M., Ritchie, H. and Ortiz-Ospina, E., 2015, 'Internet', Our World in Data (ourworldindata.org).

—, 'World population growth', 2019, Our World in Data (ourworldindata.org).

Rousseau, J.-J., 1762, *Emile, or On Education: Book I* (reprint by Heritage Books, 2019, accessed online at Perlego.com).

Rushton, D., 2020, 'London vs New York: which city has the higher average building height?', Carto (carto.com).

Sachs, J., 2020, *Ages of Globalization* (Columbia University Press).

Samans, R., et al., 2015, 'Inclusive growth and development report 2015', World Economic Forum (weforum.org).

Samora, P., 2016, 'Is this the end of slum upgrading in Brazil?', The Conversation (theconversation.com).

Sample, I., 2007, 'Drug-resistant form of plague identified', *Guardian*.

Sanchez-Paramo, C., et al., 2021, 'COVID-19 leaves a legacy of rising poverty and widening inequality', World Bank (worldbank.org).

Sanghani, N., 2019, 'Lessons from the history of globalisation', Capital Economics (capitaleconomics.com).

Schneider, B., 2018, 'CityLab University: induced demand', Bloomberg News.

Schwedel, A., et al., 2021, 'The working future: more human, not less', Bain & Company (bain.com).

Semuels, A., 2017, 'Japan might be what equality in education looks like', *The Atlantic*.

Shapiro, S., 2015, 'New species of city discovered: university city', Next City.

Shelter, 2022, 'The story of social housing' (england.shelter.org.uk).

—, 2022, 'Social housing deficit' (england.shelter.org.uk).

Sheng, A. and Geng, X., 2018, 'China is building 19 "supercity clusters"', World Economic Forum (weforum.org).

Sinclair, U., 1906, *The Jungle* (reprint by Penguin Classics, 2002).

Singapore National Parks, 2020, 'A city in nature' (nparks.gov.sg).

—, 2020, 'Annual report' (nparks.gov.sg).

Smitham, E. and Glassman, A., 2021, 'The next pandemic could come sooner and be deadlier', Center for Global Development (cgdev.org).

Song, X., et al., 2019, 'Long-term decline in intergenerational mobility in the United States since the 1850s', *Proceedings of the National Academy of Sciences*, Vol. 107, No. 1.

Spengler, O., 1922, *The Decline of the West, Volume Two* (reprint by Arktos Media Ltd, 2021).

Squitieri, A. and Altaweel, M., 2018, *Revolutionizing a World: From Small States to Universalism in the Pre-Islamic Near East* (UCL Press).

Startup Genome, 2022, 'Startup ecosystems' (startupgenome.com).

Stier, A., et al., 2020, 'COVID-19 attack rate increases with city size', Working paper.

Stoll, B., et al., 2022, 'Benchmarking European cities on creating the right conditions for zero-emission mobility', Clean Cities Campaign (cleancitiescampaign.org).

Sturge, G., 2022, 'Migration statistics', House of Commons Library (commonslibrary.parliament.uk).

Sunstein, C., 2007, 'Neither Hayek nor Habermas', *Public Choice*, Vol. 134.

Sustain Europe, 2019, 'Oslo European Green Capital 2019' (sustaineurope.com).

Swinney, P., 2019, 'Are cities bad for the environment?', BBC (bbc.com).

Tarigan, E. and Karmini, N., 2022, 'Indonesia's sinking, polluted capital is moving to new city', Associated Press.

Tempest, K., 2021, *On Connection* (Faber & Faber).

Thatcher Archive, Interview for *Woman's Own*, 23 September 1987, Margaret Thatcher Foundation (margaretthatcher.org).

Toffler, A., 1980, *The Third Wave* (Bantam Books).
Transport for London, 2022, 'Latest TfL figures show continued growth in ridership following lifting of working from home restrictions' (tfl.gov.uk).
—, 2022, 'Fares from 1 March 2022' (tfl.gov.uk).
Trust for Public Land, 2022, 'ParkScore index' (tpl.org).
Turchin, P., 2009, 'A theory for formation of large empires', *Journal of Global History*, Vol. 4, No. 2.
Turner, F., 2006, *From Counterculture to Cyberculture* (University of Chicago Press).
UN Department of Economic and Social Affairs, Population Division, 2022, 'World Population Prospects' (population.un.org).
UN Women, 2022, Ukraine and the food and fuel crisis: 4 things to know, unwomen.org
UNFAO, 2022, 'State of the world's forests' (fao.org).
United Nations, 1969, 'Growth of the world's urban and rural population, 1920–2000' (un.org).
University College London, 2017, 'Audience members' hearts beat together at the theatre' (ucl.ac.uk).
Van Alstyne, M. and Brynjolfsson, E., 1997, 'Electronic communities: global village or cyberbalkans', *Economic Theory*.
Van De Mieroop, M., 2021, *A History of Ancient Egypt* (Wiley).
Vaughan, A., 2022, '40°C heatwave may have killed 1000 people in England and Wales', *New Scientist*.
Verbruggen, R., et al., 2014, 'The networked city', in Knox, P. (ed.), *Atlas of Cities* (Princeton University Press).
Villazon, L., 2022, 'Vertical farming: why stacking crops high could be the future of agriculture', Science Focus (sciencefocus.com).
Waber, B., et al., 2014, 'Workspaces that move people', *Harvard Business Review*.
Wade, L., 2020, 'From Black Death to fatal flu, past pandemics show why people on the margins suffer most', Science.org.
Walters-Lynch, W., 2020, 'What 50 years of remote work teaches us', BBC (bbc.com).
Weale, S., 2021, 'Headteachers call for reform of school admissions to redress attainment gap', *Guardian*.
The Week, 2021, 'The rise of the superbugs: why antibiotic resistance is a "slow-moving pandemic"' (theweek.co.uk).

Wester, P., et al., 2019, *The Hindu Kush Himalaya Assessment* (Springer).

Westfall, C., 2021, 'Mental health and remote work: survey reveals 80% of workers would quit their jobs for this', Forbes (forbes.com).

White, G., 2015, 'Long commutes are awful, especially for the poor', *The Atlantic*.

Whitehouse, H., et al., 2022, 'Testing the Big Gods hypothesis with global historical data: a review and "retake"', *Religion, Brain & Behavior*.

Whyte, M., et al., 2015, 'Challenging myths about China's one-child policy', *The China Journal*, Vol. 74.

WIEGO Statistics Database (wiego.org).

Wilkerson, I., 2016, 'The long-lasting legacy of the Great Migration', *Smithsonian Magazine*.

Wilson, B., 2020, *Metropolis* (Jonathan Cape).

Wilson, E. O., 1984, *Biophilia* (Harvard University Press).

—, 2012, *The Social Conquest of the Earth* (Liveright).

Wilson, W., 1987, *The Truly Disadvantaged* (University of Chicago Press).

Winck, B., 2021, '4 charts show how fast everyone is flocking back to big cities', *Business Insider*.

Wirth, L., 1938, 'Urbanism as a way of life', *American Journal of Sociology*.

Witthaus,, J., 2022, 'From Los Angeles to New York, underused office buildings become apartments amid housing shortage', CoStar (costar.com).

Woetzel,, J., et al., 2018, 'Smart cities: digital solutions for a more livable future', McKinsey Global Institute (mckinsey.com).

Wood, M., 2020, *The Story of China* (Simon & Schuster).

World Bank, 2022, 'Four decades of poverty reduction in China' (worldbank.org).

World Bank Data, 2018, 'Population living in slums (% of urban population)' (data.worldbank.org).

—, 2022, 'Air transport, passengers carried' (data.worldbank.org).

—, 2022, 'Research and development expenditure (% of GDP)' (data .worldbank.org).

World Bank, UNESCO and UNICEF, 2021, 'The state of the global education crisis: A path to recovery' (worldbank.org).

World Meteorological Organization, 2021, *Atlas of Mortality and Economic Losses from Weather, Climate and Water Extremes (1970–2019)* (wmo.int).
World Population Review, 2022, 'World City Populations 2022' (worldpopulationreview.com).
World Population Review, 2023, 'Dhaka Population 2023' (worldpopulationreview.com).
Yeung, P., 2022, '"It's a beautiful thing": how one Paris district rediscovered conviviality', *Guardian.*
Yokoyama, M., 1991, 'Abortion policy in Japan: analysis from the framework of interest groups', *Kokugakuin Journal of Law and Politics*, Vol. 29, No. 1.
Young, R., 2014, 'Take bushmeat off the menu before humans are served another Ebola', The Conversation (theconversation.com).
Zhao, M., et al., 2022, 'A global dataset of annual urban extents (1992–2020) from harmonized nighttime lights', *Earth System Science Data*, Vol. 14, No. 2.

INDEX